Wily Monkeys

Wily Monkeys

Social Intelligence of Tibetan Macaques

Hideshi OGAWA
Translated by
Akie YANAGI

Kyoto University Press

TRANS PACIFIC PRESS

First published in 2006 jointly by:

Kyoto University Press
Kyodai Kaikan
15-9 Yoshida Kawara-cho
Sakyo-ku, Kyoto 606-8305, Japan
Telephone: +81-75-761-6182
Fax: +81-75-761-6190
Email: sales@kyoto-up.gr.jp
Web: http://www.kyoto-up.gr.jp

Trans Pacific Press
PO Box 120, Rosanna, Melbourne
Victoria 3084, Australia
Telephone: +61 3 9459 3021
Fax: +61 3 9457 5923
Email: info@transpacificpress.com
Web: http://www.transpacificpress.com

Distributors

Japan
Kyoto University Press
Kyodai Kaikan
15-9 Yoshida Kawara-cho
Sakyo-ku, Kyoto 606-8305
Telephone: (075) 761-6182
Fax: (075) 761-6190
Email: sales@kyoto-up.gr.jp
Web: http://www.kyoto-up.gr.jp

Asia and the Pacific
Kinokuniya Company Ltd.
Head office:
38-1 Sakuragaoka 5-chome, Setagaya-ku, Tokyo 156-8691, Japan
Phone: +81 (0)3 3439 0161
Fax: +81 (0)3 3439 0839
Email: bkimp@kinokuniya.co.jp
Web: www.kinokuniya.co.jp
Asia-Pacific office:
Kinokuniya Book Stores of Singapore Pte., Ltd.
391B Orchard Road #13-06/07/08
Ngee Ann City Tower B
Singapore 238874
Tel: +65 6276 5558
Fax: +65 6276 5570
Email: SSO@kinokuniya.co.jp

UK and Europe
Asian Studies Book Services
Franseweg 55B, 3921 DE Elst, Utrecht, The Netherlands
Telephone: +31 318 470 030
Fax: +31 318 470 073
Email: info@asianstudiesbooks.com
Web: http://www.asianstudiesbooks.com

USA and Canada
International Specialized Book Services (ISBS)
920 NE 58th Avenue, Suite 300
Portland, Oregon 97213-3786
USA
Telephone: (800) 944-6190
Fax: (503) 280-8832
Email: orders@isbs.com
Web: http://www.isbs.com

Australia and New Zealand
UNIREPS
University of New South Wales
Sydney, NSW 2052
Australia
Telephone: +61(0)2-9664-0999
Fax: +61(0)2-9664-5420
Email: info.press@unsw.edu.au
Web: http://www.unireps.com.au

ISBN 1-920901-97-3, 978-1-920901-97-4

On Translation

This book, originally titled "Tachimawaru Saru" ("Monkeys Play Their Cards Well"), was first written and published by Dr. Hideshi Ogawa in Japanese in 1999, and introduced the social relationships and behaviors of the Tibetan monkeys at Huangshan in Anhui province, China.

A few years earlier, I was given the opportunity to translate this book from Japanese into English through my graduate advisor, Dr. Carol Berman, who is also a primatologist at the State University of New York at Buffalo. She carried out her own project on the Tibetan monkeys at Huangshan and at that time was interested in knowing what the social behavior of these monkeys was like during Dr. Ogawa's study period. One day, she came up to me and asked if I understood the Japanese language, so I said "yes." Then she went on to ask me if I was interested in translating a Japanese primatology book into English for her, so she could read it. Of course, I said "yes!" From then on, I translated one chapter per week and finished translating every chapter by the end of the semester. About the time I was finishing up the translation of the book, Dr. Berman told me that Dr. Ogawa was interested in publishing the translated version of his book.

I myself, as a primatologist, am interested in the area of social intelligence of primates, and especially because of this reason, I was very pleased to be the one who translated this book from Japanese into English, and who could help introduce the social behavior of the Tibetan monkeys to English-speaking readers. I currently live in Puerto Rico, having joined a German team which is investigating the social behavior of the semi-free-ranging rhesus monkeys on Cayo Santiago—which is a species mentioned a little bit by Dr. Ogawa in this book—in relation to their paternal relationships. The goal of our team is to shed light on the question of whether rhesus monkeys recognize their paternal relationships with others and whether they behave on the basis of such recognition. My own doctoral project will follow after my participation on this team, which will also focus on the social intelligence of the juvenile and infant rhesus monkeys of the same population by examining the use of their play signals and play behavior.

Not only does this book present some interesting primatological findings about the Tibetan monkeys at Huangshan, but it will also be of interest to any scholar or student, including myself, who studies social behavior and/or social intelligence of primates, as well as other animals. I am very confident that even those who merely love animals or animal behavior will find this book interesting by reading about the many fascinating episodes that happened to the monkeys in the Yulingkeng group. Though this book may not cover all the aspects of animal research, it can also provide a great resource for the students of animal behavior in that they can learn what the actual researchers go through before, during, and after their fieldwork from this book—such as finding a research topic, recognizing and identifying the individuals, analyzing the results, etc. The readers of this book will also have the pleasure of getting to know Dr. Ogawa through his often-humorous style of writing, in which he introduces the interesting social behavior of the Tibetan monkeys. I hope that the readers of this book will enjoy getting to know the Tibetan monkeys at Huangshan and their unusual, but very fascinating, behaviors as much as I did when I read the original book.

In closing of this chapter, I would like to show my appreciations to a few people. I greatly appreciate Dr. Ogawa for giving me—me, as merely a graduate student—this great opportunity to translate his original book, "Tachimawaru Saru," into English and to participate in the publication of this book. I would like to thank Dr. Carol Berman for her constant support and encouragement in my career in primatology, introducing me to Dr. Ogawa, and giving me the opportunity to translate this book in the first place. I would also like to thank Mr. Shigekazu Takagaki at the Kyoto University Press for all the guidance he provided for me throughout the process of the publication of this book. Finally, I would like to thank and dedicate this book to my family, especially my mother, Atsuko Yanagi, and my father, Hiro Yanagi, for their unconditional love and support.

On December 22, 2005
Akie Yanagi

Foreword

Tibetan monkeys (Tibetan macagues) (*Macaca thibetana*) live at Mt. Huangshan in Anhui province, China. They are called 'Duanweihou' in Chinese. ('Duan' means short, 'wei' means tail, and 'hou' means monkey in Chinese.) They belong to the genus *Macaca*, in which Japanese monkeys (*Macaca fuscata*) are also included. Although Tibetan monkeys are bigger and fatter than Japanese monkeys, their appearance is not so different from that of Japanese monkeys. Like Japanese monkeys, Tibetan monkeys live in multi-male multi-female groups, and normally the males will transfer from one group to another, while the females will stay in their natal group and reproduce their offspring within the group. Compared to Japanese monkeys, however, Tibetan monkeys are peculiar in that they engage in some interesting social behaviors that we cannot observe in other monkeys. They are also peculiar in their social relationships among individuals in the social group. They will carry an infant to another individual and hold up the infant together to get along with each other. Males will show their penises to another male and lick the penises of other males. Males will also embrace each other while touching their penises. In agonistic interactions, females will form alliances with unrelated females as well as with their mother, daughters, and sisters. At night, all the group members will get together and huddle to sleep on the ledge of the mountain.

What do they think during their social interactions with other individuals?

I stayed at Huangshan in China several times from December 1989 to April 1993 and carried out a study on the social behavior of these Tibetan monkeys. In this book, I would like to show what kinds of social intelligence Tibetan monkeys have and how wily they are in their social situations, as well as introducing their unique social behaviors.

Monkey names and their behavior characteristics in this book

Huangmao

This was the alpha male (the highest-ranking male) in the Yulingkeng group during my first observations from 1989 to 1990. His fur was bright yellow, and that is why he was named 'Huangmao.' ('Huang' means yellow and 'mao' means fur in Chinese.) Tibetan monkeys become darker as they get older, so I think he was still young in 1989. He was usually preoccupied with mating with females.

Wushang

The second highest-ranking male in 1989–1990. Observers loved him because we believed that he had a nice personality. He wore a radio wave telemeter on his neck, which allowed observers to find the group in the mountain before 1988.

Duyanlong

Born in 1984. He was named 'Duyanlong' because one of his eyes was deformed. ('Duyanlong' means one-eyed dragon in Chinese.) My impression is that he has a simple mind. He was still young and low-ranking male in 1989, but he became the alpha male in 1990 and had great influence on the social relationships among females.

Sanxiong

Born in 1985. He was a good friend of Duyanlong's as if to be a young brother of him. He was popular among the observers because he often exhibited funny and interesting performances with other members in the group.

Zhi

She is very calm female. She had many offspring, Teng, Changlian, Chawei, Keai, and so on. She was the second highest-ranking female in 1990. However, she lost a fight with the female, Ye, on December 8, 1990, and became lower ranking after the fight.

Beihen

Born in 1986. She appeared easygoing, but also seemed like an artful woman who took advantage of innocent men for her own success. Many special incidents and unique experiences occurred among the females during her growth to maturity.

Baijamao

She was the highest ranking female in the group in 1989–1990. She looked like an old witch with so many scars on her face due to falling from a cliff multiple times. She finally died from falling from a cliff to the basin of a waterfall on November 26, 1990. People in the village ate her carcass afterwards.

Zhiming

Son of Zhi. Born in 1989. The two males, Sanxiong and Youshizhi, favored Zhiming so much that they and other males often held him when he was an infant.

Huagang

Son of Hua. Born in 1989. He was a good friend of Zhiming's.

Keai

Daughter of Zhi. Born in 1990. I should have named her after 'Zhi' as I did for 'Zhiming' to show her kin relationships with Zhi. However, I named her 'Keai' because she was so cute. ('Keai' means cute in Chinese.)

Huangmao

Duyanlong (right), Sanxiong (left)

Beihen

Huagang (right), Zhiming (left)

Wushang

Zhi

Baijamao

Keai

Contents

Chapter 4

*

Bridging Behavior

Chapter 5

*

Studying Wild Monkeys

Chapter 6

*

The Lives of the Males and Females

Chapter 7

*

Alliances among Females

Chapter 8

*

Infanticide

Chapter 9

*

Huddling at the Sleeping Site

Chapter 10

*

How Do Primates Understand the Mind and Behavior of Other Individuals?

Chapter 1

The Life of the Tibetan Monkeys at Huangshan

1. The formation of huddling clusters

It was evening, the time when Tibetan monkeys (Tibetan macagues) gather at their sleeping site. On January 9, 1990, I lived in the Huangshan mountain in China, where I was observing these Tibetan monkeys on a daily basis.

I followed the monkeys throughout the day, and conducted other important observations in the evening. Before long, Tibetan monkeys belonging to the Yulingkeng group gathered on one of the rocks of a steep cliff. On cold nights, all the members of the group came to this cliff to huddle up and sleep together on the rock. In order to closely observe the monkeys' social behavior at their small sleeping site, I sat at my usual vantage point—a rock overlooking a beautiful landscape. When I checked my watch, it was 4:22 p.m.

When I looked over the cliff and across a deep valley to the other side, a monkey was already seated on a rock protruding from the cliff. She was an adult female named Zhi, who had given birth to several offspring. Tibetan monkeys live in multi-male and multi-female groups. While the males move from one group to another during their lifetime, the females remain in their natal groups throughout their lives and bear offspring in these groups. Zhi lived in this group along with her three daughters and two sons. Her five-year-old daughter Teng, who would probably bear her first baby in the next spring, approached Zhi.

Shortly, another female, Beihen, appeared. She was three years old, and although her appearance was still that of a juvenile, she was sexually attractive to the males. She began mating last fall. An adult male, Huangmao, followed this young female, Beihen. Huangmao was the strongest male in this group. However, since he was not likely to lead the group, we did not refer to him as the "boss" or the "leader" but merely as an "alpha male," that is, the highest-ranking male who dominates over the other males. Since he was the highest-ranking male, he could intervene if another male was mating with a female and prevent them from being successful. On the other hand, he did not hesitate to mate in front of the other males. Certain females were attracted to such a strong male and would attempt to choose him as their mating partner. Would Beihen mate with Huangmao? The answer was "No." When he approached Beihen, she avoided him and retreated into a small cave. This cave is a tiny crack in a rock and can accommodate only two or three monkeys.

Another three-year-old female, Tougou, approached and groomed Huangmao, as if to say, "How about me?" Grooming is a behavior in which monkeys remove lice, their eggs, dust, and dandruff from their fur in order to keep themselves clean. Among primates, familiar individuals often sit beside each other and clean each other's fur by what is referred to as social grooming. Tougou rhythmically parted Huangmao's fur with both her hands. She combed his hair with her thumb and index finger and at times, deftly removed something from his fur. At that moment, she may have removed lice eggs that had been laid on his fur. Huangmao, who was being groomed, appeared comfortable. It is thought that social grooming makes monkeys feel calm.

Baijamao was the next to appear. She was a dignified-looking adult female—the highest-ranking female. While Tougou was grooming Huangmao, Baijamao inspected Tougou's genitalia. In order to better examine her buttocks, Baijamao made Tougou pose with her hands and feet on the ground. She then inspected Tougou's genitalia and touched it with the tip of her finger. Next, Baijamao inspected Beihen's genitalia. After completing the genital inspections of these two juvenile females, Baijamao sat down beside Beihen in the cave, as if to say, "These two are still too young." Beihen, who had avoided Huangmao by entering the cave, began to groom Baijamao faithfully. Zhi entered the cave too. Both Zhi and Baijamao were powerful individuals who played central roles in the social relationships of the Yulingkeng group.

Zhi passed Huangmao and Tougou while on her way to the rock. Huangmao, who was being groomed by Tougou, inspected Zhi's genitalia, as if to say, "Hey, wait!" However, he did not attempt to mate with her. Zhi arrived at the rock safely after having her genitalia inspected. The females, Zhi and Baijamao, usually come to the sleeping site early and sleep in this small cave. Huangmao, too, followed Zhi to the rock as if to say, "I will keep you up all night." Male primates, in fact, generally prefer mating with middle-aged females—who have experienced childbirth and raising offspring, and who can guarantee continued reproduction—rather than young females who do not have this experience. Although Huangmao appeared to have followed Zhi, I believe that he may have merely wanted to approach Beihen. However, Beihen left the cave, again avoiding Huangmao, who persistently followed her and disappeared into the woods. This was just as I had surmised.

Baijamao's daughters, Baixiang, who was not yet one year old, and one-year-old Sanjiaolian, joined Baijamao and Zhi in the cave. Since it was time for them to rest with their mother at the sleeping site, they may have joined her after having finished playing together in the woods. Another one-year-old infant, Zhi's daughter, Keai, may have been on her way to the site.

One after the other, the Tibetan monkeys belonging to the Yulingkeng group gathered at the sleeping site in the woods of Huangshan in China. The Yulingkeng group consists of 25 monkeys. Before all the members of the group fall asleep on a narrow ledge, with an area of approximately 10 m × 10 m, they display various behaviors. I patiently continued observing them through my binoculars.

A calm, adult male, Erxiong, approached and sat down at site A. Site A is a horizontal ledge on which Tibetan monkeys often form the largest sleeping huddle. Immediately below the ledge, the cliff falls almost perpendicular to the floor of a ravine. Teng arrived at the ledge and displayed a "presenting" behavior in front of Erxiong. This is a behavior by which their genitalia can be easily seen by another individual. Females do not necessarily resort to presenting behavior in order to solicit mating; however, they do so as a form of greeting. After this presenting behavior, Teng began grooming Erxiong.

Chiyan and Yanbaiban arrived next. I apologize that so many monkeys with new names have been mentioned consecutively in this book. However, in actual fact, I did not know much about these two males. They were adult males who had recently moved to this group. These new immigrant monkeys had experienced difficulties in this group because they were bullied by the resident members. They probably wanted to associate with the females and mate with them. However, they had to cope with the other males, too. When Chiyan arrived at site A, however, Erxiong had left without evincing any reaction. Chiyan was then groomed by Teng, who had been grooming Erxiong and was alone once Erxiong left. The other new immigrant male, Yanbaiban, also sat down beside Chiyan and Teng. Chiyan was fortunate enough to be groomed by a young female, Teng. However, as more members of the group gathered at the sleeping site, the new immigrants may have encountered some difficulty in determining the manner in which they ought to behave in front of the numerous resident members. I suppose they were thinking, "Let's sit here until the time comes."

Erxiong ignored the two new immigrant males and approached a four-year-old female, Feimao, who welcomed him with the presenting behavior. As Feimao performed this behavior, she placed her head on the ground, pushed her buttocks up in the air, almost standing on her hands. She raised her buttocks high in front of the male's face. Would Erxiong mate with Feimao because of her behavior? It was not that simple. The alpha male, Huangmao, had returned to the sleeping site since Beihen, as I had expected, had given him the cold shoulder. He had probably witnessed the interaction between Erxiong and Feimao. He approached Feimao and mated with her. Erxiong could not interfere with their copulation. Rather, he retreated in order to avoid them. Huangmao inspected Feimao's genitalia, mounted her, held up her buttocks, and inserted his penis. Huangmao thrust his penis several times and ejaculated. This took only a few seconds. After Huangmao dismounted, there were white sperms around Feimao's genitalia and on Huangmao's penis. Since sperms adhering to the penis can serve as a good source of food, he carefully picked and ate them. Huangmao, the alpha male, then headed toward site A. According to me, he did not go there to seek the new immigrant males, Yanbaiban and Chiyan. He approached Teng, who was sitting beside them. He might mate with Teng next. He mated with a large number of females, and I wondered whether he cared about his mating partners. However, Teng quickly avoided Huangmao and left. Huangmao sat down beside Yanbaiban as though he was thinking, "Shoot, I have to give it up for now." When Teng saw Huangmao sit down, she returned, sat down beside him, and groomed him; she appeared to be thinking, "I must groom the alpha male instead of mating with him."

Erxiong was at site B, which is located several meters away from site A. This, too, is a place where many monkeys often huddled together and slept. When Erxiong sat down, Feimao, who had just finished mating with Huangmao, arrived again, and Beihen emerged from the cave. Right beside her, an adult female, Ye, was grooming her two-year-old daughter, Eshang.

Wushang, the second-ranking adult male in this group, approached them, holding Zhiming to his belly. Zhiming, Zhi's son, was not yet a year old. He was born last spring and would soon be one year old. He often left his mother and played energetically. Although Wushang was holding him, Zhiming may not necessarily have been his son. As I have described, the alpha male, Huangmao, had attempted to mate with many different females up to this point. Both male and female Tibetan monkeys in multi-male and multi-female groups mate with multiple partners during the mating season in any given year. Therefore, it must be impossible for males to discern which babies born during the next spring are their offspring. It is probable that females, too, do not know who the fathers of their babies are.

The reason Wushang held one of the infants in the group is that male Tibetan monkeys hold babies in order to facilitate their social interactions with other males. During this year, it was a four-year-old juvenile male, Sanxiong, who had held infants most frequently. When Sanxiong noticed that Wushang was holding

Zhiming, he ran up to Wushang, bent down in front ofhim and lifted both of Zhiming's legs. Wushang then lifted Zhiming's shoulders. Zhiming was turned on his back, resembling a bridge between these two males. I refer to this behavior as "bridging behavior." The monkeys continued this strange, ritualistic behavior for a few seconds. Sanxiong then groomed Wushang, as if to say, "O. K., everything went well." Zhiming was still caught between Wushang and Sanxiong, who were facing each other. After a while, Zhiming extricated himself and left. He may have wanted to join a group of playing infants as he was bored with being caught between the two males. He ran toward a one-year-old male, Guer, and a one-year-old female, Keai, who were playing chase and escape. However, Zhiming could not join them because Sanxiong chased after him, held him, and ran to Wushang to present Zhiming to him again. Together, Wushang and Sanxiong lifted Zhiming. The bridging behavior was performed again. Sanxiong groomed Wushang once more, as if to say, "O. K., everything went well again." This time, Sanxiong carefully held Zhiming to his belly to prevent him from escaping.

A two-year-old male, Chawei, arrived. He was a good friend of Sanxiong and followed him everywhere. I expected Chawei, like Sanxiong, to present an infant to Wushang. However, Chawei opted for a different type of interaction with Sanxiong, who was his friend and almost like his brother. Chawei may have been able to relate more easily with Sanxiong than with Wushang. Chawei jumped on Sanxiong in order to perform a behavior known as mounting. Mounting is a behavior in which a male mounts another male and assumes the same posture as that during mating between a male and a female. I wondered whether Sanxiong was all right. Although he had flattered Wushang a great deal by presenting him an infant for bridging and then by grooming him, he now completely ignored Wushang and performed the mounting behavior with Chawei. When Wushang appeared to be unhappy about this, Sanxiong ran away. Wushang, too, stared at Chawei and left.

As explained earlier, although Zhiming—who was not yet a year old – tried to play with other infants, he was repeatedly caught by Sanxiong and used as a bridge between two males. During this commotion, Zhiming approached site A, where Huangmao, Yanbaiban, Chiyan, and Teng were seated. Zhiming was then caught by one of the new immigrant males, Yanbaiban. Yanbaiban happened to be sitting next to the alpha male, Huangmao; this is because Huangmao approached Teng and sat beside her while she was grooming Yanbaiban. Yanbaiban was probably in a very tense situation, and as Zhiming voluntarily walked toward him, he took advantage of a good opportunity to offer the alpha male a present. Yanbaiban held Zhiming and presented him to Huangmao. After Zhiming had been used in the bridging behavior between the males so many times, he finally joined his mother, Zhi, and huddled with her.

By 5:00 p.m., the sleeping site was crowded with numerous members of the group. Several monkeys made physical contact with each other in a cluster. I refer to this situation, in which several monkeys gather together and maintain physical contact, as a "huddling cluster."

The alpha male, Huangmao, was still wandering around between some of these huddling clusters. I did not know what provoked him, but he suddenly attacked Ye, who was resting with her daughter. Ye jumped up, cried out, and escaped. Then, Huangmao dragged Feimao out of the huddling cluster in order to inspect her genitalia. He had his way in whatever he wanted. In order to avoid Huangmao, Beihen, who was sitting beside Feimao, left silently. Huangmao did not notice this. Ye, who had been attacked by Huangmao and had escaped, huddled with her daughter, Eshang, again. One of the new immigrant males, Chiyan, hesitantly lingered beside them. The other new immigrant male, Yanbaiban, appeared to be becoming accustomed to the group and more active toward the females. Therefore, he followed the three-year-old female, Tougou. Although she was not completely mature, she had begun mating during the fall. However, since Yanbaiban was a low-ranking male in the group, he could not mate at the sleeping site where so many members had gathered. If Tougou went into the woods, he would have had an opportunity to mate with her. Hence, Yanbaiban returned and joined the huddling cluster consisting of Ye and her daughter, Eshang. I am unsure whether Yanbaiban thought, "I want to be with Tougou. But she must come back to her mother, Ye, soon. Therefore, I do not have to follow her now." He may have merely wanted to rest in a huddle that was less likely to be involved in any aggression. Another female, Canglian, also joined that huddle.

I continued watching this process and thought, "O. K., the huddling clusters are growing larger. Once the monkeys settle down in their huddles and stop moving, my observations for today will be complete." The reason I wished they would hurry and finish forming their huddles was that it was very cold. Monkeys who slept in the very center of a huddling cluster could remain warm, but I was alone. Actually, I was accompanied by my research colleagues who were observing the monkeys with me; they were shivering too. However, it would have been strange if we were to huddle together like the Tibetan monkeys.

Huangmao approached again and attempted to mate with Canglian. As a result, the huddle dispersed.

Erxiong, Feimao, Duyanlong, and Wushang were huddling together at site B. Canglian, who had finished mating with Huangmao, arrived at this huddle and joined them. The young males—Sanxiong, Guer, and Chawei—were huddling some distance away from site B. Zhi, her daughter, Teng, and her son, Zhiming,

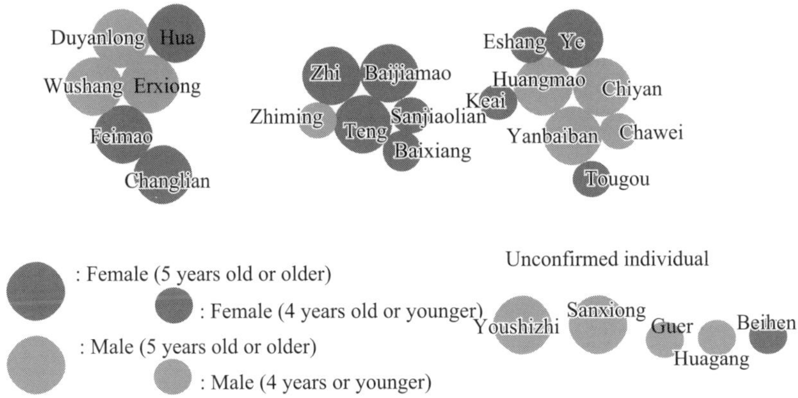

Fig. 1-1 Disposition of the huddles at the sleeping site on Jan. 19, 1990.

and Baijamao, with her daughters Sanjiaolian and Baixiang, were in the cave. They had settled down completely. At site A, those members of the group who had dispersed because of Huangmao's aggressive behavior began assembling again. Ye, her daughters Tougou and Eshang, the new immigrant males, Chiyan and Yanbaiban, and the alpha male, Huangmao, were present at the site. Two infants, Keai and Chawei, were also crouched at the fringes of the huddle. They appeared to be preparing to settle down (Fig. 1-1).

Finally, a female, Hua, arrived at the sleeping site. She was a young adult who had given birth to her first baby the previous spring. The alpha male in the Yulingkeng group, Huangmao, who was preoccupied with mating, immediately leaped out of the huddle and mated with Hua. Although the other members of the group witnessed this, they did not move and remained in their huddles.

The Yulingkeng group included another male, whom I have not mentioned yet. He was a low-ranking adult male named Youshizhi. Since all the other members had already secured their positions in the huddle, I wondered whether it was too late for him to find a huddle he could join. However, I noticed that Youshizhi was holding an infant—Hua's son, Huagang—to his belly. I expected that Youshizhi would attempt to join the huddle by presenting the infant to the alpha male, Huangmao, and by performing the bridging behavior and grooming Huangmao. However, when he arrived at the sleeping site, Huangmao was in the process of mating with Hua. Youshizhi thoughtlessly performed the bridging behavior with Yanbaiban, who was right beside Huangmao. Huangmao finished mating, and attacked Youshizhi. The huddling clusters, which had, at length, been formed, all dispersed at 5.34 p.m.

Since it had grown too dark, no more observations were possible that day. My research colleagues and I hurriedly descended the mountain to the village of

Fuxi.

2. The beginning of my study

I began formulating my academic plans during the fall of 1989, while I was a graduate student at the Primate Research Institute, Kyoto University, Inuyama City, Japan. I had to decide where I would carry out the field observations of primates. During that summer, I visited several places, such as Yakushima Island, Koushima Island, Takasakiyama, Kinkazan Island, and Arashiyama, where other primatologists had been conducting fieldwork on the ecology and social behavior of Japanese monkeys (*Macaca fuscata*). I was undecided on which of these places I ought to select to conduct my research.

Currently, young researchers who have recently graduated occasionally have the opportunity to initiate long-term studies in places such as the Mahale National Park in Tanzania, where long-term observations of wild chimpanzees have been conducted. I wished to begin my fieldwork abroad and study wild animals as soon as possible. I was drawn to the world of wild animals that I had watched every week on TV programs such as "The Wild Kingdom," when I was a little boy. However, at that time (the 1980s), it was necessary for graduate students who had majored in primatology to study Japanese monkeys for the first two years of their doctoral course in order to train themselves as researchers. If the students were interested in studying wild primates in countries other than Japan, they were finally provided with opportunities to pursue such interests during the latter three years of their doctoral course.

One day, at the Primate Research Institute, I found a hand-written note from Dr. Tetsuro Matsuzawa on my desk. Dr. Matsuzawa is famous for his study of a female chimpanzee named "Ai." Ai is well known because she has received language training and displays a high level of intelligence. The note ran as follows: "Dr. Kazuo Wada is seeking a student to accompany him on a preliminary study of golden snub-nosed monkeys (*Rhinopithecus roxellana*), conducted by the Northwest University in Shanxi province, China." Dr. Wada was a member of the Primate Research Institute and had studied wild monkeys in China. I immediately descended the stairs in order to meet him; however, my classmate, Naoki Agetsuma, had pre-empted me, and Dr. Wada had already decided to take him on the preliminary study of golden snub-nosed monkeys. However, he noticed that I was disappointed, and asked, "Would you like to go to the field where I am currently conducting research on Tibetan monkeys? Tibetan monkeys are pretty interesting monkeys. We can observe their social behavior, and therefore, I believe that the study of these monkeys will be interesting." I immediately came to a decision and said, "Yes, I will. I will study Tibetan monkeys." In truth, I found this proposition very attractive because of the place— "Tibet." Although I soon realized that the field in which the Tibetan monkeys were to be studied is at Huangshan in Anhui Province and is very far from Tibet, I had already made up my mind and was not inclined to reconsider.

Usually, when scientists begin their research, they do not decide on their field of study as promptly as I did. Before beginning research on a new topic, they review previous studies related to that topic. They assess which hypotheses have already been proposed and which ideas are novel, and come to the conclusion that while certain things are known, others remain unknown. After completing this process, they draft a study plan regarding the methods of research and the animal that will be suitable for their observations or experiments. However, to date, little is known about the ecology and social behavior of Tibetan monkeys. Nevertheless, I was extremely fortunate to be in a situation in which I could actually observe these monkeys. One month later, I flew to China to study the Tibetan monkeys.

Thus, I commenced my observations of the Tibetan monkeys. I first arrived at the field at Huangshan on December 3, 1989. From then on, up to April 28, 1993, I occasionally boarded at Huangshan in order to conduct my observations and, at times, in Japan in order to carry out data analyses. In this manner, I continued studying the Tibetan monkeys.

3. Visiting Huangshan

The field where the study was conducted, Huangshan, in Anhui province, China, was originally discovered by Mr. Chengpei Xiong from the Hefei Xiaoyaojin Zoo (currently known as the Hefei Wild Animal and Plant Protection Center). Mr. Chengpei Xiong designated Huangshan as a suitable place for long-term field study after conducting an investigation of the distribution of all the primate species in Anhui province. Since 1973, he has paid several visits to Huangshan and has established cordial relations with the people of the village of Fuxi at Huangshan. On the other hand, Dr. Kazuo Wada, who was interested in the study of wild animals in China, had begun a joint research project with Anhui University in 1985. They formed a primate research group comprising Dr. Kazuo Wada, Prof. Qishan Wang from Anhui University, Mr. Jinhua Li, a graduate student of Anhui University, and Mr. Chengpei Xiong from the Hefei Xiaoyaojin Zoo. In 1989, another graduate student from Anhui university, Mr. Ming Li, and I joined this research group. Since then, we have studied the ecology, society, and social behavior of the Tibetan monkeys. Based on our convenience and study topics, each of us took turns at living in the village of Fuxi at Huangshan, and we continued our observations of the Tibetan monkeys with the assistance of Mr. Xinming Cheng and the other residents of the village. (1)

Conducting research on wild animals often requires laborious work before the researcher actually enters the habitat of the animals. Although Huangshan is not too far from Japan, a certain amount of time elapsed before I commenced my observations of the Tibetan monkeys at Huangshan. I first left Japan for Shanghai, and then traveled from Shanghai to Anhui University in Hefei, the capital city of the Anhui province. As preparation for my study, I completed all the necessary paperwork at Anhui University and finally headed for Huangshan. At present, China has a well-developed infrastructure, and highways and bridges have been built rapidly. However, when I first visited Huangshan in 1989, it was a full day's journey away from Hefei. I covered the approximately 400 km between Hefei and Huangshan in a van, which Dr. Wada had brought from Japan. On the way to Huangshan, I crossed the Changjiang River by ferry and then ate my lunch. My journey finally ended in the evening, when I arrived at the village of Zhaixi, which is located below Huangshan. From the village of Zhaixi, I followed a mountain path along the Fuxi River for 40 minutes and managed to arrive at the village of Fuxi just before sunset (Fig. 1–2). I lived in this village and carried out my research there.

I recommend that the best route to take in order to visit Huangshan and see the Tibetan monkeys in the village of Fuxi is to fly to Tunxi and take a microbus or a taxi to Huangshan, which is a few hours away. I traveled from Tunxi to Huangshan by a long-distance bus. This bus had three seats on the left side and two on the right, with a narrow aisle running down the center. The seats were so small that it was very difficult to sit in them for any considerable amount of time. But as I had no option, I endured the discomfort and focused on the view through the window. As the bus passed the village of Zhaixi on its way to the gates of Huangshan, I informed the bus driver that I wished to alight and got off the bus. In order to visit the village of Fuxi and see the monkeys after enjoying the sights at Huangshan, a visitor can take a 10-minute ride in a taxi from Tangkou—located below Huangshan—to the village of Zhaixi, which is the entrance to the village of Fuxi. Huangshan has been designated a world heritage site according to the World Heritage Treaty, and it is one of the best sightseeing areas in China. However, the package tours available to foreign visitors, which arrange for several days in China, do not always include a visit to Huangshan. Yet, it is a very famous place that is known to all Chinese people. I will now introduce Huangshan as a sightseeing spot. Huangshan comprises not merely one mountain but 72 peaks, and covers an area of 154 km^2. Huangshan City is located at longitude 118° 11" east and latitude 30° 29" north. This latitude is

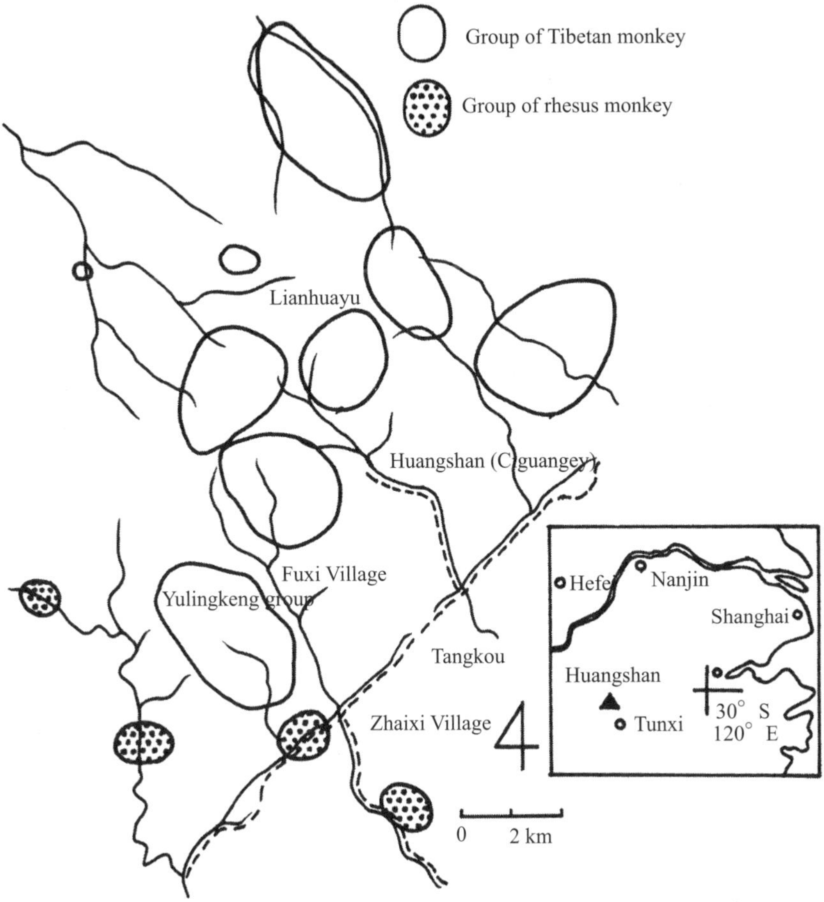

Fig. 1–2 Distribution of Tibetan monkeys at Huangshan.
A revised version of Fig. 5, Wada et al., 1986.

almost the same as that of Yakushima Island in Japan. The annual average temperature is 15 degrees Centigrade and the annual rainfall is 1,500 mm, as measured by the meteorological observatory that is located 200 m above sea level. There is a hot spring at the foot of the mountain. While the hot spring is enjoyable, it is different from a Japanese-style hot spring. From this point, it is possible to climb the mountain. The climb from the front entrance, Ciguangey, leads to the Baihai area, which is the high point of sightseeing tours. Yet another path—the route from Yungusi Temple to the Baihai area—involves taking a ropeway. Nevertheless, I recommend the walking route from Ciguangey to the Baihai area via Yupinglou, and Yingesong (the Welcoming Pine Tree). This

Photo. 1-1 Landscape of Huangshan, "Monkey Gazing at the Sea."

route threads through the mountain ridges and includes the stone steps of the mountain path. It also affords a view of rocks under the cover of pine trees and several strange rocks, one of which is known as "Feilaishi (the Flying from Afar Stone)." A visitor with enough time and good health can climb Tiandufeng Peak or Lianhuafeng Peak, which are about 1,800 m tall. The following day, it is possible to visit Qingliangtai, which is easily accessible from Baihai hotel in the Baihai area. From there, it is possible to see the rock known as the "Houziguanhai (Monkey Gazing at the Sea)," which is at the summit of the cliff (Photo. 1-1). When this rock is covered by a sea of clouds, it appears as though a monkey is sitting down and looking at the sea. I have heard that a long time ago, there was once a small rock that resembled a tail attached to this rock.

However, it is almost impossible to encounter Tibetan monkeys or other large mammals when you are walking on these sightseeing routes at Huangshan, because the monkeys are disturbed by the people walking on these paths, and pine trees and rocks do not provide food for monkeys. Tibetan monkeys can only be seen in forests of broad-leaved trees. I will now proceed to the village of Fuxi.

Photo. 1-2 Tibetan monkeys (*Macaca thibetana*).
Body weight: 15 kg (male), 10 kg (female)
Body size: 62 cm (male), 52 cm (female)

4. Monkeys belonging to the genus *Macaca*

Before describing the life of the Tibetan monkeys at Huangshan, I will describe the general features of Tibetan monkeys and other monkeys in the same genus as well.

Tibetan monkeys (*Macaca thibetana*) belong to the order Primates in the class Mammalia. The order Primates includes monkeys, apes, and humans (*Homo sapiens*). In China, which is home to 17 primate species, monkeys are referred to as "Houzi" in Chinese. According to a poem written by Bai Li, gibbons, too, lived on the north shore of the Changjiang River a long time ago. At present, however, the habitat of primates as well as that of other large mammals has shrunk considerably on account of deforestation. Previously, monkeys were also hunted for food and medicine. At present, all primates in China fall under the category of protected species—Tibetan monkeys are a protected species of the second level, and golden snub-tailed monkeys and gibbons are protected species of the first level.

Tibetan monkeys belong to the genus *Macaca* (macaque in English) in the order Primates. The Japanese monkeys that the people of Japan are familiar with, also belong to this group. Therefore, macaques are the type of monkeys that most Japanese people would typically imagine.

On a slightly different note, I am often asked whether humans evolved from monkeys. The correct answer, however, is that humans evolved from the common ancestor of monkeys and humans and that the monkeys of today have also evolved from the common ancestor of monkeys and humans. Thus, if the statement, "Humans evolved from monkeys" implies that, for example, Japanese monkeys are our ancestors, this assumption is incorrect. The ancestor of humans and apes and the ancestor of Japanese monkeys parted ways a long time ago. The group comprising the ancestors of humans and apes, which had been prosperous at one point in time, declined and came under pressure from new species of monkeys such as macaques and baboons. From this dwindling group of humans and apes, it was humans alone who, in recent times, were in fact able to expand to a population of approximately 6,000,000,000 individuals.

It is held that macaques originated from the areas surrounding the Mediterranean Sea. Fossils of macaques dating from approximately 6 million years ago have been discovered in these areas. Barbary monkeys (*Macaca sylvanus*), which also belong to the genus *Macaca*, currently inhabit North Africa. However, monkeys belonging to the genus *Macaca* have diversified in Asia. They inhabited various regions in Asia and branched into various species approximately 2 million years ago. As a result, macaques can be divided into 16–19 species, all of which live in Asia, except for Barbary monkeys, which live in North Africa.(2) Five species of macaques inhabit China: Tibetan monkeys, rhesus monkeys (*Macaca mulatta*), Assamese monkeys (*Macaca assamensis*), stump-tailed monkeys (*Macaca arctoides*), and pig-tailed monkeys (*Macaca nemestrina*). (3)

5. Monkeys at Huangshan

Two species of the genus *Macaca* inhabit Huangshan: Tibetan monkeys and rhesus monkeys. (4)

Rhesus monkeys are smaller than Japanese monkeys and have brownish tails that curl upwards. In Chinese, they are referred to as "Mihou." Rhesus monkeys are widely distributed over Southeast Asia, China, and India. In terms of their distribution and the size of their population, they are one of the most successful species in the genus *Macaca*. Some of them obstinately inhabit the scattered and impoverished environment of very small forests in cultivated areas. Therefore, it is understandable that they sometimes eat the crops and damage the harvests in China. Street performers in China often conduct shows with rhesus monkeys. In some countries, the monkeys are strays, in a manner similar to stray dogs. Thus, rhesus monkeys are closely associated with human life.

Tibetan monkeys, on the other hand, are larger than Japanese monkeys and have short, thick bodies with short tails. In Chinese, they are referred to as "Duanweihou" (Short-tailed monkeys). The environment and the areas inhabited by Tibetan monkeys are more limited than those of rhesus monkeys. The habitat of the Tibetan monkeys is restricted to China. In this sense, they are a species peculiar to China. The areas they are distributed in range from Anhui province to Sichuan province; they cover both sides of the Changjian River, including Hubai province, Guizhou province, Hunan province, Jiangxi province, Zhejiang province, and Fujian province. However, the habitats of the Tibetan monkeys are scattered within the distributional areas of rhesus monkeys. In Africa and South America, two or more different primate species sometimes form a mixed group and forage together. However, Tibetan monkeys and rhesus monkeys do not even inhabit the same forests although their distributional areas overlap. For example, while the Tibetan monkeys at Huangshan inhabit forests that are at higher than 600 m above sea level, rhesus monkeys inhabit the lower areas. On the mountains that are not inhabited by Tibetan monkeys, rhesus monkeys sometimes live in areas that are more than 600 m above sea level.

Huangshan comprises forests of evergreen broad-leaved trees, among which bamboo thickets are scattered, at less than 600 m above sea level. There are mixed forests of evergreen and deciduous trees between 600 and 1,200 m, and forests of deciduous broad-leaved trees are located at altitudes above 1,200 m. Tibetan monkeys usually forage in the mixed broad-leaved forests of evergreen and deciduous trees that comprise their habitat.[(4)] In these forests, there exist several species of trees that are closely related to those in Japan. Tibetan monkeys, like Japanese monkeys, feed on buds, flowers, young leaves, and bamboo shoots during the spring, and on fruits and seeds (chestnuts and acorns) during the fall. During the winter, they chew on winter buds and peeled barks of trees, or look for seeds buried under fallen leaves. I rarely noticed them eat insects. However, this may be due to the fact that I observed them mainly during the winter. During the spring, I sometimes saw them catch and eat live insects. I often began my observations during the late fall and completed them at the beginning of spring; therefore, I regret that I was unable to accumulate quantitative data on their food sources throughout the year.

According to a study conducted by Dr. Wada et al., Huangshan was inhabited by nine groups of Tibetan monkeys[(4)] (Fig. 1–2). All the groups inhabited areas consisting of craggy cliffs. As mentioned at the beginning of this book, the monkeys use the ledges of the steep cliffs as their sleeping sites.

On my arrival at Huangshan, one of these groups of Tibetan monkeys, known as the "Yulingkeng group," was selected as the target of our study and was closely observed. This group comprises approximately 30 individuals. The Yulingkeng group has been provisioned since 1985 in order to facilitate observation. After being provisioned, each individual monkey was identified. A valley containing a waterfall that is approximately 5 m in height is located on the steep slope of the mountain. A provisioning site was established downstream from the waterfall in an open space on the banks of a river. Corn, which is usually food for pigs and is known as "Yumi'' in Chinese, was scattered at the site four times in a day, that is, at 7:00 a.m., 10:00 a.m., 1:00 p.m., and 3:00 p.m.

Upon arrival at the village of Fuxi, the life of the Tibetan monkeys in this Yulingkeng group can be observed. However, I request the reader's patience, and in the next chapter, I will describe the problems confronted by this group of monkeys.

Chapter 2

Monkeys Play Their Cards Well

1. Evolution of social intelligence

It is often held that monkeys are intelligent. However, the assessment of this "intelligence" poses an extremely difficult problem. We may believe that monkeys are intelligent merely because monkeys and humans are phylogenetically close and share several common characteristics. However, we can evaluate intelligence on the basis of the size of an animal's brain. Primates have a larger brain than other animals. Animals that have a large body require and do, in fact, have large brains. Yet, even if we exclude this factor, primates still have a large brain relative to their body size. Further, in primates, the cerebrum (the cerebral cortex) and the frontal lobe occupy a large proportion of the brain. This region of the brain is what is known as the "thinking" region; it processes information obtained from the external world through visual stimuli and is responsible for making choices. In actuality, the "intelligence" of monkeys includes their ability to solve various problems in flexible ways that are not limited to particular situations.

How have primates evolved this ability?

The reason that certain animals today have specific physical or behavioral characteristics is that in the environment of these animals, the individuals that had such characteristics were able to produce more offspring than individuals that did not. It is crucial for wild animals to procure food on a daily basis and, in case they have predators, to avoid becoming their prey. In other words, it is important for wild animals to utilize their environment efficiently and survive without being taken advantage of by other animals. Reproduction is another vital factor. It is necessary for male mammals that perpetuate by means of sexual reproduction to copulate successfully and as frequently as possible, and for females to successfully raise their babies to adulthood. Therefore, the more offspring they produce, the more will be the individuals in the next generation that will inherit desirable characteristics from their parents, and the genes that produce a particular characteristic will be disseminated in the gene pool of the species. The number of offspring that an individual produces during its lifetime is referred to as "fitness." This fitness denotes the number of offspring that will mature to a reproductive age. It is believed that the reason animals possess certain characteristics is that the individuals that possess these characteristics have a higher fitness. This process by which animals evolve is known as "natural selection."

Genes are also disseminated in the gene pool of the species if the individual possessing the genes can make its related individuals such as siblings produce a large number of offspring, instead of producing its own offspring. This process is known as "kin selection." In this case, it is important for the index of "inclusive fitness" to be high. The inclusive fitness includes the increase or the decrease in the fitness of an animal's relatives based on the interaction between that animal and its relatives, as well as the animal's own fitness.

In general, we can easily assume that animals that possess certain additional abilities are more likely to survive and reproduce than other animals that do not possess these abilities. However, the possession of certain abilities comes at a cost. For example, digestive organs that are capable of digesting grass fibers are advantageous to animals that live in the savannas. However, possessing such digestive organs and the actual execution of their activities requires energy and time. Similarly, large brains that enable animals to solve problems flexibly may be beneficial in certain respects; however, such brains always require a considerable amount of energy and oxygen. If the benefits of possessing certain characteristics are greater than all the costs of possessing them, then those characteristics evolve. Larger brains may have afforded greater benefits to primates. Monkeys have evolved large brains because, with their ability to solve problems flexibly, they have taken advantage of other animals and their external environment, and they have avoided being taken advantage of by other animals. This ability may have contributed to their survival and reproduction.

Several experiments have measured the ability of primates to solve problems flexibly; this ability is not limited to a particular situation. W. Keller's classic experiment is famous among these. In this experiment, he constructed problems that can be solved through the efficient manipulation of objects. For example, a chimpanzee was able to reach a banana hung from the ceiling if (s)he stacked boxes or connected sticks. Did these animals and our ancestors inhabit an environment in which this type of ability facilitated the procurement of food? Does such an environment account for our intelligence?

However, the environment of animals does not merely comprise the physical environment, such as the climate, and the ecological environment, which consists of the threat posed by predators or the availability of food resources. In the case of animals that live in a group, the group itself also constitutes an environment. Group-living animals create their environment along with the other members of the group. This type of environment is known as the "social environment." It is important to note that primates often exhibit the ability to solve problems flexibly during social interactions in their social environment. The intelligence that animals use in the course of their interactions with others is known as "social intelligence." This ability differs from the ability with which a solitary individual obtains food through the efficient utilization of its physical and ecological surroundings. In addition, there has been support for the hypothesis that primate intelligence has evolved as an adaptation in order to solve various social problems in a complex social environment.[1][2] The fact that primates who live in a large group are more likely to have larger cerebrums also supports this hypothesis.[3] [4] Given that a large group gives rise to a more complex social environment for the individuals in that group, the large cerebrums of primates may have evolved in such a complex social environment.

However, the reason that great apes and humans, and humans in particular, have remarkably large brains cannot be explained by this hypothesis alone.

Nevertheless, if we consider the way in which we use our brains everyday, we must be impressed by the fact that we often reflect on and are concerned about our social relationships with the people around us. Humans are, in fact, social animals. For example, a person who unexpectedly loses his job would be concerned about how he will earn money from the next day onward, while in the train on his way home. He may casually think, "I'm hungry. What's for dinner? Oh, there is a good-looking woman sitting in front of me. " He may later begin thinking about his unsuccessful social relationships in the workplace, or about the way in which he ought to break the news to his family. Currently, a large number of people who take entrance examinations to enter certain universities or companies are evaluated on the basis of interviews or other tests that measure their social abilities, as well as their memory power or their ability to think logically, all of which can be assessed by written tests. The various aspects of the importance of social intelligence have attracted greater attention than various other human abilities. This is reasonable because primates, including humans, have evolved their intelligence primarily in order to behave more efficiently in social situations.

2. Problems encountered in group life

Group life entails various problems; this is evident on reflecting upon ourselves. Why do humans and monkeys live in groups? This is because, as opposed to a solitary life, each individual obtains greater advantages than disadvantages from group living. One of the reasons monkeys live in groups is that this reduces the risk of predation. However, in cases in which they live in an environment that has few predators, or in which they do not run the risk of being eaten by predators because of their relatively large physical stature, there is another reason that accounts for group living. This reason is food. While cows can easily digest grass fibers found anywhere, most monkeys cannot digest large quantities of grass or leaves. They prefer fruits to other parts of plants since fruits are easy to digest and provide a lot of nutrition. However, fruits, unlike leaves, are not evenly distributed but are dispersed in patches. Monkeys may find many fruits in a big tree during certain seasons, but may be unable to find any in another tree. In order to obtain such food, it is practical for them to unite and protect a good food patch together. Protecting their food resources is one of the important reasons that monkeys form groups. (5)

In order to ensure that both monkeys and their offspring are able to obtain food efficiently, it is preferable for them to form a large group composed of individuals that are related; this will enable them to compete with other groups. However, if the size of the group becomes excessively large, another problem, namely, the decrease of the food resource per individual, will arise. The members of the same group also have to compete with one another over limited food resources. If resources are broadly defined as factors that animals use to increase their fitness, then resources include individuals of the opposite sex of the same species with whom the animals reproduce. Some monkeys may form a group in order to defend individuals of the opposite sex against other groups. At the same time, monkeys belonging to the same group compete with one another for mating. Monkeys that belong to the same group are both competitors over resources and cooperators who defend themselves against other groups.

This ambivalent situation encountered by the monkeys creates a delicate environment. How should they behave in such a situation? If their interests coincide, they can communicate with one another and select the option that benefits everyone. On the other hand, if their interests are entirely conflicting, there is no room for communication with one another. In this case, it would be practical for them to compete with one another in order to obtain the resources. However, in the case of monkeys, it is possible for them to modify their behavior such that they remain content even though they have to compete for

Fig. 2-1 Payoff matrix of the hawk and dove, prisoner's dilemma, and chicken games.

Table A. Hawk and dove game

Player A ＼ Player B	Dove	Hawk
Dove	(V/2) ＼ (V/2)	0 ＼ (V − c)
Hawk	(V−c) ＼ 0	{(V/2)−C} ＼ {(V/2)−C}

In each cell, the first variable denotes the payoff obtained by player A, and the second, the payoff obtained by player B.
When two doves encounter each other, they share V, and each obtains (V/2). When a dove and a hawk encounter each other, the dove obtains 0 and the hawk incurs a small cost, c, and obtains V. When two hawks encounter each other, each incurs a large cost, C, and obtains V by a probability of 50%.

Table B. Two players game.

Player A ＼ Player B	Cooperation	Defection
Cooperation	R ＼ R	S ＼ T
Defection	T ＼ S	P ＼ P

R: Reward for mutual cooperation. T: Temptation to defect. S: Sucker's payoff. P: Punishment for mutual defection.
The game is called a prisoner's dilemma game if $T > R > P > S$, and $2R > T + S$ in Table B.
The game is called a chicken game if $T > R > S > P$ in Table B.

Table C. An example of a prisoner's dilemma game.

Player A ＼ Player B	Cooperation	Defection
Cooperation	3 ＼ 3	0 ＼ 5
Defection	5 ＼ 0	1 ＼ 1

When $V = 6$, $c = 1$, and $C = 2$ in Table A.

Table D. An example of a chicken game.

Player A ＼ Player B	Cooperation	Defection
Cooperation	3 ＼ 3	0 ＼ 5
Defection	5 ＼ 0	−1 ＼ −1

When $V = 6$, $c = 1$, and $C = 4$ in Table A.

resources. What is the optimum form of behavior in this case? Generally, communication is effective in obtaining the desired outcome. However, there exist certain cases in which communication cannot resolve all the issues.

In order to obtain the best choice in the competition for resources, we will now consider a basic model known as the "hawk and dove game." The hawk and dove game presented in Fig. 2-1 presents a simple situation in which two players compete over a certain resource and select one of two behaviors (strategies). However, since obtaining the resource entails certain costs, this cannot be a zero-sum game in which the interests of the players are entirely in conflict. The players that select the "hawk" type of behavior to obtain the resource regardless of the amount of energy and time expended or the extent of the injuries they sustain. Those that select the "dove" type of behavior obtain only half the resource by simply sharing it, provided the opponent is also a dove. Now, if the value of the resource is (V), the players obtain (V/2). However, a dove will forfeit everything if its opponent is a hawk. The hawk sustains a small cost (c) in order to establish that it is a hawk, and obtains the entire resource (V). However, if the hawk competes with another hawk, both will fight for the resource and sustain a considerable cost (C), and the hawk that wins by a probability of 50% will obtain the resource (V). Let us assume that these players encounter each other at random and compete over the resource. In this case, it may be preferable to compromise and leave the entire resource to the competitor rather than fight and spend all one's energies to procure the resource.

In this game, if the value of the resource (V) and the costs of obtaining it, (c) and (C), are scored, we can construct, for example, Table C, shown in Fig. 2-1. The game that includes this type of measurement of scores is known as a "prisoner's dilemma game. " In this type of game, irrespective of the opponent's strategy, a player obtains a higher score if (s)he selects the strategy, "defection," which is a "hawk" in the hawk and dove game.

However, in the case of scoring as presented in Table D, the hawk will compete for the resource even when the other competitor is a hawk, and the cost of the competition exceeds the average value of the resource. The game involving this type of measurement of scores is known as a "chicken game." If the group has a larger proportion of doves, a hawk will obtain a high score since it frequently encounters doves and secures all the resources. On the other hand, if the group has a greater proportion of hawks, the score of the hawk will be lower than that of the dove because of the frequent encounters between hawks. In this case, the score of the dove will be higher than that of the hawk although the dove will grant the entire resource to the hawk when it encounters the hawk. In the case of scoring as presented in Table D, if doves (cooperation) constituted 1/3 of the group and hawks (defection) constituted the remainder 2/3, the average score of these two strategists would become even. Assuming that the strategist who obtains a higher score produces more offspring, and if doves are in a proportion greater than the one mentioned above, then the number of hawks will increase in the next generation as the conditions are more favorable to them. On the other hand, if hawks are in a proportion greater than the one mentioned above, then the number of doves will increase in the next generation as the conditions favor the doves. As a result, the proportion will return to the level at which the benefits for both are even. From the evolutionary perspective, this proportion indicates a stable situation. In short, Table D presents the situation in which the choice (strategy) that is favorable to the players depends on which strategist exists in the group.

On the contrary, in the case of a prisoner's dilemma game, as presented in Table C, the hawk always obtains greater benefits. This is because even if two hawks encounter each other and compete for the resource, they can still obtain, on an average, a higher score than they would by merely forfeiting the resource. The prisoner's dilemma game has been closely studied as a model according to which cooperative and altruistic behaviors have evolved. Altruistic behavior refers to a behavior that increases the fitness of others and reduces one's own fitness. Let us consider the following case. A player may select either altruistic behavior or selfish behavior. If altruistic behavior is selected, the player's score will decrease by 1 point and the opponent's score will increase by 4 points. If the player chooses selfish behavior, his or her score will increase by 1 point and the opponent's score will decrease by 1 point. If we calculate four sets of results in which the player and the opponent select each option once, the scores amount to a prisoner's dilemma game. In this game, it is undoubtedly more beneficial to choose selfish behavior (defection). It is natural that the behavior that lowers one's own fitness does not prevail in the group unless the opponent is kin-related. However, even under these circumstances, if, for example, one has to compete with the same competitor repeatedly, there exists a better option than being always selfish (defection in Table C and hawk in Table A). I will not examine the details at this juncture; however, it is observed that the "Tit for Tat" strategy is fairly effective in a case in which the player repeatedly competes with an opponent in the prisoner's dilemma game.[(6)] In the Tit for Tat strategy, players repeat the behavior selected by the competitor in the previous competitive interaction. According to computer simulations, individuals that adopt this strategy are the ones that prevail in the group. This indicates the possibility that altruistic behavior has evolved on the basis of mutual profits by reciprocity. Further, the computer simulations indicate that in a situation wherein the players occasionally mistake each other's cooperation for defection, as often occurs in the real world, the following two strategies can be adopted by a large proportion of the members of the group for an extended period of time: (1) the Generous Tit for Tat strategy in which the players usually adopt the Tit for Tat strategy, but sometimes excuse the opponent's defection, and (2) the win-stay and lose-shift strategy in which the players repeat the same behavior if they obtain the best or the second-best points among all the four sets, and adopt a different behavior if they obtain the worst or second-worst points.[(7)]

3. Predicting and manipulating the behavior of other monkeys

How would humans select their behavior if they were to play a prisoner's dilemma game or a chicken game? I sometimes make the students in my biology class play these games.(8) In a game consisting of only one round, the students are made to guess whether their competitor will cooperate with them or defect to them on the basis of factors such as the competitor's personality. When the students were repeatedly matched with the same competitors, for example in 10-round bouts, they managed not only to guess their competitor's choice but also to elicit the competitor's cooperation by recalling his or her past choices. This is because obtaining the competitor's cooperation and maintaining mutual cooperation by making a good choice oneself is the key to securing high points.

We can assume that the monkeys, too, are in a situation similar to that of the students playing the above mentioned games. The monkeys simultaneously cooperate with the other members of the group and compete with one another for resources. Therefore, if they are familiar with the characteristics of their competitor, they can effectively predict his or her behavior. Further, if they are able to predict their competitor's subsequent behavior judging from previous interactions, then they can behave in a manner that is favorable to themselves. In other words, choosing their own behavior by predicting the behavioral choices of other individuals will afford considerable benefits to the animals that live in groups. Several group-living animals, including primates, do in fact change their behavior depending on the behavior of others. In addition, if they can predict the responses that their behavior will elicit, they will be able to determine their behavior in advance and manipulate the behavior of other individuals. Even when they cannot compel other individuals to comply with them, they can alter the mental condition of other individuals by teaching them something or deceiving them, and can thereby indirectly manipulate their behavior.

Are monkeys capable of predicting the behavior of other individuals according to their past behavioral patterns and manipulating these individuals on the basis of these predictions? Takao Fushimi constructed the following interactive situation in his experiment and examined the ability of the sender of information.(9) I will present the procedures of the experiment in a simple manner. Two monkeys seated face-to-face are allowed to push one of the two buttons placed in front of them. First, the sender of the information pushes either button A or B. The receiver watches this process. Assume that the receiver pushes the button P; in this case, food will be delivered to the sender as a reward. If the receiver continues pushing Q, the sender is helpless. However, if the receiver is trained to push P when the sender pushes A, and to push Q when the sender pushes B, the sender should ideally push A in order to make the receiver push P. Further, if the receiver changes his response and pushes Q when the sender pushes A, the sender must select the button by guessing which button the receiver will push in response. The result of this experiment showed that in order to manipulate the receiver's behavior, Japanese monkeys determined their own behavior according to their predictions regarding the receiver's response to the information they sent.

In their lives in the wild, once monkeys came to possess these abilities, namely, the ability to manipulate the behavior of others and the counter ability to predict the behavior of others in order to ensure that they would not be manipulated in an undesirable manner, these abilities simultaneously reinforced one another and evolved. [(10) (11)]

4. Social interactions involving three individuals

Social intelligence is also exhibited in a situation in which three or more individuals are involved, as well as in the interactions between two individuals.

For example, it is probably advantageous for group-living primates to have a reliable partner as an ally when they compete with others. Let us consider the following coalition formation game. Assume that each player, A, B, C, and D, has 4, 3, 2, and 1 points, respectively, of the ability to compete for the resource. If two individuals compete for the resource, the one with the higher competing ability will win. The individual who anticipates a loss may forfeit the resource to another individual who has a higher ability. It is natural that the prerogative over the resource assumes the order A, B, C, and D. However, if, for example, B and C form a coalition, their total ability becomes 5 (3 + 2) and these individuals will be able to secure a victory over A. The prerogative over the resource can therefore assume the order B, C, A, and D. The formation of such a coalition may result in the prerogative over the resource that is actually observed among monkeys, in other words, what is referred to as "dominance ranks." Since each monkey must pursue his or her profits, the monkey who has the highest ability does not necessarily establish a successful coalition. The coalition is a crucial factor that determines whether the competitors in the group secure the resource.

If the monkeys are aware of the individual social relationships within a group, they can take advantage of a third individual in the course of their interaction with another individual. As an example of intelligent behavioral choices in social interactions involving three individuals, let us consider the interactions between twin Japanese monkeys and their mother, which I observed for my graduate thesis project. [(12) (13)]

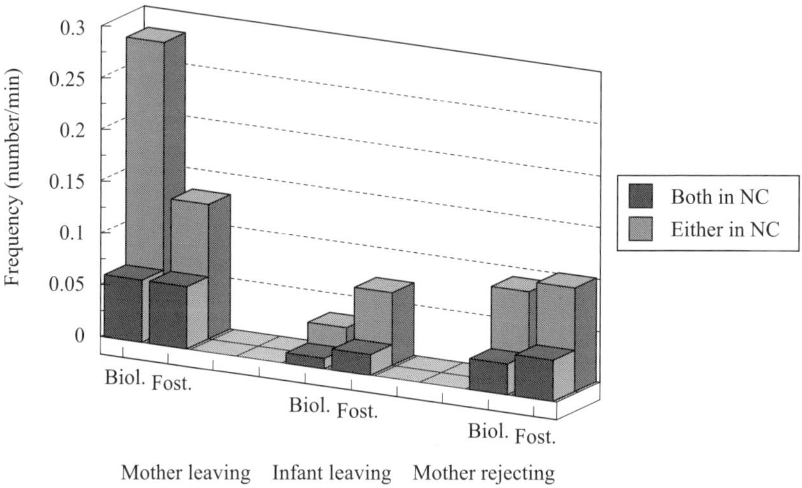

Fig. 2–2 Interactions between twins and their mother of Japanese monkeys.
Both in NC: Both the infants were in nipple contact with the mother. Either in NC: Only one of the infants was in nipple contact with the mother. Mother leaving: The mother left the infant. Infant leaving: The infant left the mother. Mother rejecting: The mother pushed the infant away from the nipple. Biol.: Biological offspring; Fost.: Foster offspring.

My graduate thesis was on twin Japanese monkeys belonging to the Arashiyama F group in the Iwatayama Natural Park (currently known as the Arashiyama Monkey Park, Iwatayama) in Arashiyama, Kyoto, Japan. One of the twin monkeys was adopted. However, since the female monkey, Oppress70, who fostered this baby when he was very young, did not differentiate between him from her own offspring, I would like to refer to these two monkeys as "twins." Feeding two babies at the same time was a challenging task for the mother. Fortunately, since the Japanese monkeys at Arashiyama are provisioned, Oppress70 was able to provide sufficient nutrition to her two infants. However, when an infant sucks on the mother's nipples for breast milk, there are constraints imposed on the time that the mother herself can devote to feeding and other activities. Therefore, the mother of these twins found a good solution, that is, she was less likely to remove the infants from her nipples or refuse nipple contact while two of them were feeding at the same time, as compared with when only one of them was feeding (Fig. 2–2). When both infants were in nipple contact at the same time, the mother could minimize the time constraints on her own activities as opposed to when each infant was in nipple contact at two different periods of time. The mother manipulated the behavior of the twins so as to reduce her nursing time. On the other hand, the infants would approach their mother when one of them was already in nipple contact. Further, when both of them were in nipple contact, they were unlikely to be refused, and thus, were less likely to leave their mother than when only one of them was with her and was more likely to be refused (Fig. 2–2). The behavior of the twin infants may have been a result of the mother's manipulation. In addition, this behavior also enabled them to adjust to her behavior, allowing them to obtain more nursing time for themselves. Since female Japanese monkeys usually have one baby at a time, this rare occurrence, requiring the mother to raise twins, cannot be entirely explained by a genetically programmed behavioral pattern. This example indicates that Japanese monkeys demonstrate the ability to change their behavioral patterns according to their interaction with others, even in interactions involving three individuals.

The society of primates is a complex environment. As indicated by the above example, individuals should respond to and choose their own behavior according to the situation in such a complex environment. This makes the interactions among individuals increasingly complex. Complex interactions among individuals may have further accelerated the development of intelligence in the social context. That is, primates may have evolved their intelligence in order to enable them to behave in a manner advantageous to themselves in complex interactions with others of the same species, who are simultaneously competitors and cooperators for resources. This type of intelligence is also referred to as "Machiavellian intelligence."(3) Upon a close observation of their behavior, the manner in which monkeys regulate their social interactions is clearly evident.

5. The mind of the monkey

The fact that monkeys efficiently regulate their social interactions does not imply that they are capable of predicting the behavior of other individuals in the same way that humans can. Humans attempt to predict the behavior of other people by imagining how their minds work. We think, "If I do this, they will think like this and act like this." However, we do not know for certain whether animals other than humans also understand what other individuals think. I described the function and the evolutionary factors responsible for social intelligence. However, the mental mechanisms of how and in what sense animals understand their society or the minds of other individuals have recently become a point of controversy.

When I began my study of Tibetan monkeys, I left for China without a detailed research plan. However, this does not mean that I did not have an ambitious plan in mind. In fact, while on the airplane on my way to China, I was thinking "I want to look as deep as possible into the mind of Tibetan monkeys.

What do they usually think about in their daily lives, and how? How is it similar to or different from our ways of thinking?"

Assume that you and your dog, "Koro", are walking along the banks of a river in the evening. You are watching the sunset beyond the bank. Although you and Koro are walking in the same direction, your experience of the world around you is quite different from Koro's experience of the world around him. Koro does not perceive the colors of the sunset that you see or share your impression of the beauty of the sunset. Instead, Koro hears sounds that you cannot hear and smells odors that you cannot smell.

Imagine the case of a chimpanzee, "Ai." Ai will be able to perceive colors in a similar manner to that of humans. Phylogenetically, chimpanzees are the animals that are closest to humans. From the evolutionary point of view, chimpanzees and humans parted ways from their common ancestor approximately 500 million years ago. The DNA sequence of chimpanzees differs from that of humans by only 2%. Since their sense organs that relay the information of light or sound and the functions of their brain that process such information are very similar to those of humans, there is very little difference between both species in terms of the information processing of colors and simple figures.(14) However, chimpanzees, too, do not perceive the world in exactly the same way as humans. It is possible to observe differences between these two species in high-level information processing, such as the recognition of individual faces. For example, it appears to be easier for chimpanzees to recognize the faces of other chimpanzees rather than those of humans. Chimpanzees and humans encounter different types of difficulties in the recognition of tilted or upside down faces. Studies have recently been conducted on the higher levels of various information processing abilities, such as the categorization of objects or the conception of numbers, and thus, the manner in which chimpanzees understand the world has gradually been revealed.

Nevertheless, we are unaware of what animals usually feel, enjoy, are worried about, and think about in their daily lives. The answer to such questions remains unknown. I wonder what and how the dog, Koro, the chimpanzee, Ai, and the Tibetan monkey, Huangmao, think about their lives. Among the matters concerning the manner in which monkeys usually think about their lives, the social lives of monkeys are of particular interest to me. It is also interesting to study topics such as the preferences of monkeys. For example, "Ai usually eats a banana first and then an apple. However, she prefers bananas to apples." Yet, topics such as the social interactions of monkeys hold a greater interest for me. For example, "Ai usually fights with X and they do not get along with each other. Nevertheless, she actually likes X better than Y." This is because as a human, I am concerned about my own social life, and I believe that nonhuman primates, too, have evolved their intelligence as a result of the happiness and anxiety they experience in their social lives.

While I do not prefer primates to other animals, the animal that is best suited to providing the answer to my questions is one that lives in a group and allows me to actually observe the social interactions within the group. This is why I have chosen primates as the subject of my study.

What is the concrete evidence that primates possess social intelligence, and how do they use this intelligence in their daily lives? Unlike humans, nonhuman primates do not use language. My goal is to reveal the minds of nonhuman primates who do not use language, and to inform the readers of this book of their minds on behalf of them. In order to achieve this goal, my first step was a careful observation of their behavior. I would like to pay special attention to their complex social interactions, that is, those involving three or more individuals, their social intelligence, which allows them to efficiently regulate these interactions, and the workings of their mind. At the same time, I would like the reader to enjoy reading about the interesting social life of the Tibetan monkeys, which is considerably different from that of Japanese monkeys.

Chapter 3

Observations of the Monkeys

1. The large number of male Tibetan monkeys in the group

I first arrived at the village of Fuxi at Huangshan on December 3, 1989. Mr. Xinming Cheng and the other villagers have assisted me in my research through the years. They had surveyed the mountain and had already identified the areas inhabited by the Tibetan monkeys belonging to the Yulingkeng group. The following evening, I was informed that the Tibetan monkeys were eating corn at the provisioning site in a valley. This meant that I would finally be able to observe them the following morning.

At 7:00 a.m., I left the village and began ascending the mountain; this climb took approximately an hour. When I arrived at an observational hut, which was constructed of wood, Mr. Xinming Cheng informed me that the monkeys had just arrived at the provisioning site. I hurried to the site with my binoculars and a small notepad from my backpack. The Tibetan monkeys were gathering near the river in the valley. I was impressed by the fact that I was provided with an excellent opportunity to make direct observations of the Tibetan monkeys, whose social behavior was still unknown to the world. I began observing them and recording my observations. One of the first tasks at hand was the identification of each individual in the group.

I was immediately impressed by the large number of males in the group. I noticed a considerable number of large adult males at the provisioning site. When I later took Jinhua Li to the Arashiyama Monkey Park in Kyoto during his visit to Japan, he was astonished by the small number of males and the large number of females in a group of Japanese monkeys. When he saw the Japanese monkeys eating wheat grain in an open space for provisioning, he exclaimed, "All these Japanese monkeys are females. Where are the males?" Unlike in the case of Japanese monkeys, it was natural for the group of Tibetan monkeys to include a large number of males.

In the Yulingkeng group, there were small and slender males too; these males were juveniles.

Among Japanese monkeys, the juvenile males usually linger at the periphery of the group. For instance, even when the park managers of the Arashiyama Monkey Park provide wheat grains at the provisioning site where tourists can observe the monkeys, a majority of these low-ranking juvenile males are unable to reach the center of the area and eat the grains. Therefore, after all the high-ranking males and females leave for the mountain in the evening, the park managers give these males some sweet potatoes.

Like the Japanese monkeys in Arashiyama, the juvenile male Tibetan monkeys also did not get the chance to enter the center of the provisioning site andate the corn that accidentally fell to the sides of the provisioning site or downstream in the valley outside the site. However, after the high-ranking monkeys had finished eating most of the corn within 30 minutes, these juvenile males that had remained at the periphery of the group displayed an interesting behavior that cannot be observed in Japanese monkeys. When the group members returned to the woods, we, too, made our way through the bushes and sometimes climbed over the cliffs to follow them. There are certain skills that one should possess in order to follow wild Japanese monkeys without losing sight of them. Females or high-ranking males should be followed. If a juvenile male is followed, the observer will somehow be separated from the other group members. Therefore, once the observer loses sight of the juvenile male, it is no longer possible to catch up with the group. However, this is not the case for Tibetan monkeys. Regardless of how long I observed the position of each Tibetan monkey, I did not recognize the social structure, indicated by double circles, that can be observed among Japanese monkeys: high-ranking males, females, and infants form the central part of the group, while juvenile males remain at the periphery of the group. Juvenile male Tibetan monkeys often have affiliative interactions with adult males.

There were more surprises in store. The Tibetan monkeys engaged in grooming on the rocks, they lay on the branches, and tried to find acorns on the ground. In addition to recording the behavior of individuals, primatologists usually record the proximity of the group members every few minutes. I decided to take note of the individuals who were within 5 m of a certain other individual. The individuals within this distance of the alpha male, Huangmao, were Zhi, Ye, Feimao, Youshizhi, and Canglian. Wushang, Erxiong, and Duyanlong were also present, as were Teng and several infants. Keeping track of all the monkeys was indeed a daunting task. A few of them sat close enough to touch each other, and groomed other monkeys. Another cluster, in which other monkeys were also resting together, was located within 2–3 m of this cluster. I believed that there was very little proximity between individual Tibetan monkeys.

Humans, too, unconsciously adjust their proximity to other people. For example, when we talk with another person, we regulate our distance from that person, depending on the situation and our relationship with the person. This adjustment of proximity differs among individuals and cultures.(1) Some people prefer to stand extremely close to people they may not be intimate with. Are Tibetan monkeys like these people?

Male Japanese monkeys tend to maintain a distance from other males while they are feeding or resting in the woods. Among the females as well, low-ranking females attempt to stay away from high-ranking females, especially while they are feeding. A gathering of females necessarily comprises kin-related females such as mother-daughter pairs and sisters. However, the Tibetan monkeys in front of me were very different from the Japanese monkeys I had previously observed. While they were resting on rocks, they often formed groups that included two or more adult males. These groups often included monkeys of various ages and both sexes, such as adult males, adult females, juvenile males, and infants. Unrelated females, too, often groomed each other. In general, it appeared that the Tibetan monkeys had frequent interactions with multiple and various individuals. As my observations progressed, I understood that this is the way in which the society of Tibetan monkeys functions.

2. Marking infant monkeys for identification

The Tibetan monkeys assembled together in groups. The juvenile males, too, frequently interacted with the other group members. Therefore, I was able to identify the individuals in the group quickly and easily.

On the first day of my observations itself, I was able to identify all the adult males in the group. The highest-ranking male (the alpha male) was Huangmao. He was still young and had brightly-colored hair. The second-ranking male was Wushang. He had lost some of his front teeth, as suggested by his name, Wushang, which means missing upper teeth. He appeared to be much older than Huangmao. The successive males according to the descending order of their ranks were Erxiong and Youshizhi. The latter had eyes high up on his face; this made him resemble the Metron aliens from a TV show called "Ultra Seven." He could be easily identified by the index finger of his right hand, which he could not bend. Duyanlong and Sanxiong were the juvenile males. The former had lost his left eye when he was attacked by a hawk as an infant. Sanxiong was one year younger than Duyanlong, and was very energetic. This made him easily identifiable to the observers.

Both Baijamao and Zhi were adult females and had several offspring. Baijamao was the highest-ranking among all the females and Zhi was the second highest. They must have appeared rather intimidating to the young monkeys. The other females were Ye, Hua, Teng, Feimao, and Canglian (in descending order of their ages).

At this juncture, the infants had not yet been properly identified. For example, the Yulingkeng group included an individual named "Guer." Guer means "orphan" in Chinese. He was given this name because it was believed that his mother had died. However, after close observations of his social interactions, we judged Ye to be his mother; she must have given birth to Guer during the period when the Yulingkeng group was not under observation. Within three days from the commencement of my observations, I was able to identify and name all the females and males, including the individuals that were not yet full-grown adults. However, I still faced a challenging task, namely, I wanted to identify all the members of the group, including the infants born that spring.

For my graduate thesis, I observed a total of 300 Japanese monkeys belonging to two groups in Arashiyama. For my study, however, I had to identify only the twins and their mother. Since one of the twins was adopted and did not resemble the other, I could easily differentiate between them.

Immediately before I left for China upon becoming a graduate student, I visited Arashiyama again in order to study the behavior of playmates among infant Japanese monkeys. I wanted to study the manner in which they understand the world and the individuals around them. To this end, it was advisable to gather data on the playmates, whose combinations were changed frequently. This was difficult at that point since I had to identify all the infants that were playing in this area. I was unable to identify them merely on the basis of their faces or physical features. Fortunately, the tourists in the Arashiyama Monkey Park enjoyed feeding peanuts to these monkeys through windows covered with a wire mesh. The tourists were in the observation house and the monkeys were outside. Since the monkeys climbed onto the wire mesh while they waited for the peanuts, I crept up behind them with a brush and hair dye and marked the infants with it. Thus, I was able to mark all the infants with a black spot of hair dye on their backs. In fact, some visitors wondered whether they were spotted monkeys

Unfortunately, I could not use this method at Huangshan since the monkeys were not very accustomed to people. Neither the researchers nor their assistants fed the monkeys; only the assistants fed them corn at a specific time and place. It appeared to be an impossible task to mark the infants even by using a long stick with a brush attached to the end of it.

There was one method of attracting them without using any food. Tibetan monkeys approached humans whenever they urinated. They liked to lick the leaves or rocks that were wetted by urine as they may have wanted the salt in the urine. Although I could have urinated and waited for the infants to arrive, I was not very comfortable with this method. More important, although we were not feeding them, we did not consider it advisable to accustom them to this type of practice. Therefore, we decided to urinate only in the river. At first, the monkeys waited for me to urinate, and when they saw me urinating into the river, they looked disappointed as if to say "What a waste!" After a certain point, however, they lost interest and stopped waiting. At times, Duyanlong approached me and licked the sweat off my skin. When he did this while I was conducting my observations, I wanted to drive him away. However, if he were to get provoked and make loud noises, I would have had to interrupt my observations. While I was tolerating this, Duyanlong took advantage of the opportunity and approached until he was only 5 cm away from my face, and finally licked my face. However, this incident occurred a few years after the commencement of my study in 1989.

Finally, in order to mark the infants, I borrowed a toy water gun from Dr. Wada. I marked the infant monkeys 10 days after I had begun my observations. I put hair dye in the water gun and crept up behind the infants in the woods. I tried to aim at them from a distance of approximately 2 m. However, they fled as soon as I shot them with the gun—they were very ingenious at escaping. Once I had failed, they remembered that a black substance would shoot out of the gun whenever I approached them with it. Therefore, they did not appear to be excessively cautious but rather, approached me out of curiosity. However, they managed to evade me whenever I shot at them with the hair dye. They would then approach me again and watch me from the tree branches. It seemed that they were playing a game with me. As I chased them and climbed up and down the rocks of the cliffs, I accidentally dropped the water gun and broke it. I had failed in this exercise completely. However, I observed one of the infants through my binoculars and noticed a slight black dot on his head. He was a zero-year-old male, the son of Hua, and I named him "Huagang." Another zero-year-old male, Zhi's son, was not marked by the hair dye. I named him "Zhiming." A female, a daughter of Baijamao, was born that spring, and I named her "Baixiang." After two weeks, the black dot on Huagang's head had disappeared. However, during those two weeks, I had managed to identify the peculiarities of his body and face. Once I identified those features, I could recognize him easily without the mark.

The following year, I returned with a new toy water gun from Japan. I had planned to mark the infants that were born in 1990 and were one year younger than Zhiming. However, my marking project was unsuccessful again this year. This water gun sounded an alarm when I pulled the trigger, and all the infants fled as soon as they heard it. When I had purchased the gun at a toy store, I had been informed that it was extremely popular, and I now understood why – because of this alarm. Although I regretted it, there was nothing I could do about it. That year, while I was alone on the mountain of Huangshan in China, I persisted in my attempts to shoot at the infants with the water gun that unfortunately, kept sounding an alarm. The infant monkeys enjoyed playing with and teasing me, and jumping between the branches of the trees.

3. Composition of the Yulingkeng group

Once I had identified the monkeys, I was able to understand the composition of the Yulingkeng group in 1989 (Table 3-1).

In this book, I use the word "offspring" to refer to the son or daughter of a particular female. I use the word "infant" to refer to individuals that are three years old or younger. (Both are referred to as "Kodomo" in Japanese.) Since the babies of Tibetan monkeys are usually born during the spring, the individuals I refer to as two-year-olds during the fall and winter are, in fact, two years and some months old (approximately between 6 and 10 months). In the case of the observations conducted during the spring, the individuals that were born in the previous spring are actually less than one year old, unless they have already had a birthday; however, I refer to these infants as one-year-olds.

Table 3-2 presents the transitions in the age/sex composition of the Yulingkeng group from 1989 to 1993. The group was provisioned in 1985. Although there were a large number of deaths in 1988, the population recovered, and when I began my observations in December 1989, the group consisted of 25 individuals. By the fall of 1992, this number had further increased to 42.

Females from other groups never immigrated into the study group, and all the females remained in their natal groups. Figure 3-1 presents the matrilineage of the Yulingkeng group. The males, on the other hand, shifted between groups. Some of the males from among those at Huangshan immigrated into the study group, and others disappeared leaving no trace and were considered to have emigrated from the study group they originally belonged to (Fig. 3-2).

4. Observational methods

I will now outline the observational and recording methods used to study the

Table 3–1 Age/sex composition of the Yulingkeng group in 1989.

Female					Male				
Name	Age	Age class	Rank	Remarks	Name	Age	Age class	Rank	Remarks
Baijamao	8+	Adult	1	Multiparous	Huangmao	8+	Adult	1	Natal or Immigrant before 1985
Zhi	8+	Adult	2	Multiparous	Wushang	8+	Adult	2	Immigrant before 1985
Ye	8+	Adult	5	Multiparous	Erxiong	8+	Adult	3	Immigrant in 1987
Hua	6	Adult	4	Single parous	Youshizhi	8+	Adult	4	Immigrant in 1988
Teng	5	Adult	3	Nullparous	Yanbaiban	8+	Adult	5	Immigrant in 1989
					Chiyan	8+	Adult	6	Immigrant in 1989
Feimao	4	Juvenile			Duyanlong	5	Juvenile		Natal
Canglian	4	Juvenile			Sanxiong	4	Juvenile		Natal
Tougou	3	Infant							
Beihen	3	Infant							
Eshang	2	Infant			Chawei	2	Infant		Natal
Sanjiaolian	1	Infant			Guer	1	Infant		Natal
Keai	1	Infant			Zhiming	0	Infant		Natal
Baixiang	0	Infant			Huagang	0	Infant		Natal

Table 3–2 Changes in the age/sex composition of the Yulingkeng group.

Study period		1989		1990		1991		1992	
Age class		Male	Female	Male	Female	Male	Female	Male	Female
Adult	(Male: ≧ 8 yrs) (Female: ≧ 5 yrs)	6	5	5	7	5	8	9	9
Sub-adult	(Male: 6–7 yrs)	0	-	1	-	2	-	1	-
Juvenile	(Male: 4–5 yrs) (Female: 4 yrs)	2	2	1	2	6	1	5	2
Infant	(3 yrs)	0	2	1	1	4	2	2	1
	(2 yrs)	1	1	1	2	2	1	1	4
	(1 yr)	1	2	2	1	1	4	0	3
	(6 months < and < 1 yr)	2	1	1	4	0	3	4	1
Total		12	13	12	17	20	19	22	20

Study period: 1989=Dec. 5, 1989 – Jan. 21, 1990 and March 8, 1990 – April 20, 1990; 1990=Nov. 18, 1990 – Jan. 11, 1991 and Mar. 30, 1991 – April 23,1991; 1991=Dec. 5, 1991 – April 26, 1992; 1992=Sept. 16, 1992 – Nov. 4, 1992. From 1989 to 1992, 16 babies were born, 14 males immigrated into the group, and 7 males emigrated from the group. In 1990, 1 adult female died. The number of babies (<6 months) is not shown. In 1991, 1 adult, 5 juvenile, and 3 infant males immigrated into the group. From among these immigrant males, 2 juveniles and 1 infant emigrated from the group, and 1 infant male died in 1991.

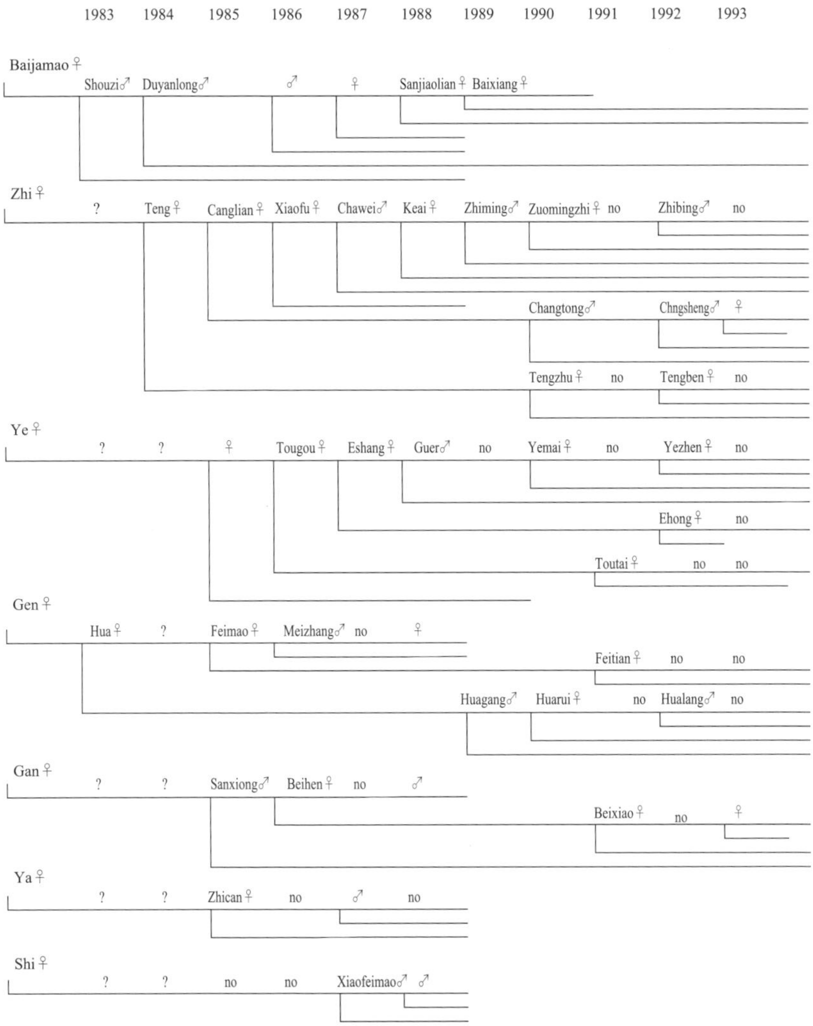

Fig. 3-1 Matrilineage of the Yulingkeng group.

monkeys. I could observe almost all the individuals in the group at the provisioning site. I strove to record any social behavior in as much detail as possible. While observing animal behavior, it is important to record the situation exactly as it is, without any preconceptions. However, it is impossible to record everything we see. It is also impossible to avoid the biases introduced by the researcher's goals. Therefore, it is necessary to prepare a list of events or states that are to be recorded, and when these actually occur, they must always be noted down according to the list.

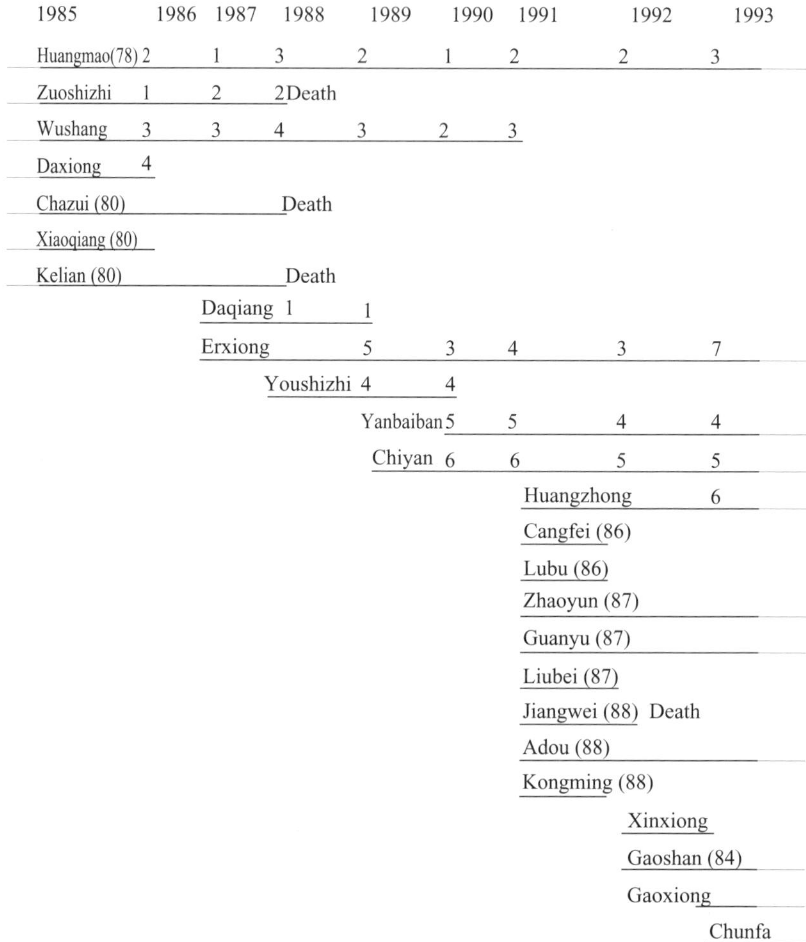

Fig. 3–2 The movements of the males of the Yulingkeng group.
Birth years of young males are estimated and shown in parentheses. Some young males who were resident in the group in 1985 may be natal of the group. Other natal males are listed in Fig. 3–1. Dominance ranks among males are shown on the lines. The natal male, Duyanlong, was the first ranking male in 1990–1992. The natal males, Sanxiong and Duyanlong, were the first and second ranking males in 1992–1993.

After they had eaten all the corn, the monkeys returned to the woods, and I followed them. While the monkeys were resting, moving, or eating, I usually chose one individual as a focal animal and recorded the behavioral events performed by this animal, its behavioral state, the social behavior that the other individuals directed toward the focal animal, and the time or the duration of each behavior. I also took note of the individuals around the focal animal at certain minute intervals. This method is known as focal animal sampling.[(2)(3)] It is the most common method used to observe primates, and allows the researcher to determine the frequency with which the focal individual performs a particular behavior with another individual.

However, if the researcher concentrates excessively on one particular individual, (s)he may fail to observe the larger picture. In addition, this method is inadequate in cases in which the researcher encounters a situation involving three or more individuals. Previously, primatologists used the focal animal sampling method and divided the interactions into dyadic ones between the focal animal and another individual even when analyzing complex interactions. However, if we do not use observations or analyses to compensate for the flaw in this method, we will be unable to entirely grasp the situations in primate society involving these complex interactions. Therefore, when the Tibetan monkeys assembled at their sleeping site in the evening, I recorded the behavior and disposition of each individual in the group.

Regardless of how systematically we attempt to record behavior, we will sometimes encounter unexpected events. It is necessary to be flexible, especially when the event is extremely rare. It is necessary to change observational plans and continue recording the event regardless of the situation. Any encounter between the observer and the primate is an once-in-a-lifetime opportunity. The Japanese tea ceremony has the teaching that "we must treasure every meeting, for it will never recur" ("Ichigo-Ichie" in Japanese). In fact, on my very first evening, I witnessed an unexpected event after the last provisioning. An adult female, Baijamao, fell off the cliff.

Observational case: The event occurred on the evening of the first day

Baijamao fell off a cliff that was more than 10 m high. She was lying on her belly and was completely motionless. The males may have believed that we had harmed her. They threatened us by loud vocalizations and expressions, and by banging on the ground. However, since we remained quiet, they climbed up the cliff and retreated into the woods. The only individual who remained in the area was Wushang. He approached Baijamao and groomed her. I was extremely moved by his actions. There was a reason he remained with her. Wushang had formed a consortship with Baijamao (This type of relationship will be discussed in greater detail in Chapter 6.). I thought that Baijamao had instantaneously died as a result of the fall. However, 10 minutes after the incident, she moved her face. She was bleeding from her right eye. Thirty minutes after the incident, she lifted herself off the ground, limped, and left the spot with Wushang. We were extremely relieved. I looked at the provisioning site and noticed someone there. It was the juvenile male, Duyanlong. Despite this horrible incident, he was still eating the corn stuck between the rocks.

Chapter 4

Bridging Behavior

1. Males interact with other males using their penises

Following my observations on the first day, I observed various events occurring in the Yulingkeng group almost everyday. However, this book would be interminable if I attempted to describe every event. Since I found male social behavior extremely amusing and interesting when I began observing the Tibetan monkeys, I would like to introduce it in this chapter. When Dr. Wada began his observations of Tibetan monkeys and presented these observations to primatologists, they were intrigued by the behaviors of the male Tibetan monkeys. Male Tibetan monkeys frequently lick the penises of other males, lick each other's penis, and embrace by touching each other's penis in various postures everyday (Fig. 4–1).

Penis sucking

Among the various social behaviors of male Tibetan monkeys, the most unique is the behavior in which a male licks (sucks) another male's penis. The male whose penis is sucked assumes a posture in which he lies down on his back, and the male who performs the action of sucking the penis leans toward him and puts his head between the legs of the first male. He then moves his lips and sucks the tip of the penis. The male toward whom this behavior is directed often chatters.

Males sometimes hold each other's body and suck each other's penis in the posture that is referred to as the "sixty-nine" position (Photo. 4–1).

Embracing

Embracing is also a frequently observed behavior, in which males embrace each other face-to-face and rock each other's body as though they are wrestling.

Fig. 4–1 Male social behavior.

Photo. 4–1 Males sucking each other's penises.

Photo. 4–2 Males embracing.

One or both of the males often stretch their hand(s) to the genital area and touch the penis (Photo. 4–2).

Mounting

Male Tibetan monkeys, like other primates, also engage in mounting behavior. One male mounts the back of another and assumes the same posture as when a male mates with a female.

Male Japanese monkeys, too, exhibit this behavior. I recall observing this mounting behavior among Japanese monkeys in Takasakiyama, in Oita, Japan. During my participation in a population census of these monkeys, I counted the number of Japanese monkeys that approached the provisioning site from the mountain in the morning. The mounting behavior I observed appeared to be a kind of solemn ceremony. A dominant male approached, silently mounted a subordinate male, and then, just as silently left.

The mounting behavior of male Tibetan monkeys is more animated than that of Japanese monkeys. Usually, both the males involved in this behavior chatter. Sometimes, as soon as one male mounts another, a third male hears their vocalizations, approaches, and immediately mounts one of them. I once observed a male mounting another male that had already mounted another male. This was like "Placing a child turtle on an adult turtle and placing a grandchild turtle on the child turtle." (This is a phrase in a Japanese song which describes the sight of several turtles being stacked over each other.) Unfortunately, I was

Photo. 4–3 Males performing mounting.

unable to photograph these males. Male Tibetan monkeys may not engage in mounting to confirm their dominant-subordinate relationships by way of a dominant male mounting a subordinate male. There was no statistically significant difference between the frequency of the mounting performed by a dominant male and that by a subordinate male. I often observed a dominant male exhibiting his buttocks to a subordinate male, looking back at him, and inviting him by raising his hand to his buttocks. The subordinate male responded to this and mounted the dominant male. Both males then chattered cheerfully (Photo. 4–3).

Presenting

When a subordinate male approaches a dominant male, he sometimes exhibits his buttocks to the dominant male before sitting down beside him. This is similar to a female exhibiting her buttocks or her genitalia to a male. In this case, the dominant male does not mount the subordinate male.

Penis showing

A subordinate male that approaches a dominant male not only exhibits his buttocks to the dominant male (presenting) but also exhibits his penis by raising his hind limbs. This posture is similar to that assumed by a dog while urinating on an electric pole. It is certain that no male Tibetan monkey would perform any action that would provoke the dominant males. The dominant male often

Photo. 4–4 Penis-sucking behavior.

responds to this penis-showing behavior by lightly touching the penis or the area around the buttocks of the subordinate male. This response relieves the subordinate male and he appears to be at ease while sitting beside the dominant male.

It is a mystery why Tibetan monkeys use their penises to such an extent in their social behavior. No individual touches or sucks his own penis. This is one of their social behavioral patterns. The penises of the males toward whom the sucking behavior is directed sometimes undergo an erection; however, they never ejaculate.

Incidentally, I am sometimes asked, "Is it true that once monkeys are taught to masturbate, they persist in this behavior all the time?" The people who ask this question are probably referring to monkeys that are kept as pets since wild monkeys cannot be taught to masturbate. This question disgusts me, and I usually reply, "It is human males who continue masturbating once they learn how, although they may not do it all the time but at least every other day."

Infant Tibetan monkeys repeatedly perform affiliative social behavior with other infants. Juvenile males perform such behavior with other juvenile males as well as adult males. In the case of penis-sucking behavior between an adult male and an infant or between two infants, the infant sometimes jumps onto his partner's face in order to have his penis sucked (Photo. 4–4). In penis-sucking behavior between an adult male and an infant, the adult male usually lifts up the

Photo. 4–5 Penis-sucking behavior.

infant and sucks his penis. The adult male may sometimes turn the infant upside down (Photo. 4–5). In this way, his face does not have to be covered by the infant's upper body.

Adult males or females perform genital-sucking behavior not only with male infants but also with female infants. An adult monkey licks the genitalia of a female infant, and a female infant occasionally licks the penis of a male infant. However, I have not observed genital sucking between an adult male and an adult female, between an adult female and an adult female, or between two female infants. Bonobos (*Pan paniscus*), which are referred to as pygmy chimpanzees or Bilia in the local language, are known to perform genital-rubbing behavior between two females. There are certain similarities and differences between Bonobos and Tibetan monkeys in the performance of this type of behavior, which if compared, may prove to be interesting.

Why do Tibetan monkeys engage in this behavior? Embracing and penis sucking are never aggressive in nature although the two individuals engaged in this behavior appear excited and they grapple with each other while making loud vocalizations. These are rather friendly behaviors. They appear to be types of greeting behavior, and their function may be to reduce the tension between two males. The frequent and various social behaviors among males may be the result of the crowded society these monkeys live in, which reduces the distance between individuals. These behaviors may also be related to the fact that many adult males live in the same group as a certain number of adult females.

In order to maximize the production of offspring, there exists competition for individuals of the opposite sex among animals that reproduce by sexual reproduction. In general, competition among males for females is more intense than that among females for males. An obvious instance of competition is usually observed when males compete with other males in order to mate with as many females as possible. One reason for this can be described as follows.

Sexual reproduction was originally evolved as the process by which creatures exchange genes with other individuals and alter their own genes in order to resist bacteria and viruses that continuously and rapidly change by way of mutations. Gene exchange was combined with asexual reproduction, and the system of sexual reproduction was brought into existence. Through sexual reproduction, gametes were formed by meiosis and were united to produce an individual of the next generation. This is the first stage in the evolution of sexual reproduction.

In the second stage, the gametes were divided into two polar types. Ova are one of the types of gametes. An ovum contains a considerable amount of nutrition that is used by the offspring during the initial stages of its life. Sperms are the other type of gamete. A sperm does not contain as much nutrition as an ovum, and as a result, all the nutrition is confined to the ovum. Sperms, however, are light and easily motile.

In the third stage, certain creatures evolved into males and females. The males exclusively produce sperms, and the females exclusively produce ova. The species that include both males and females are considered to consist of two types of individuals that adopt different reproductive strategies. The strategy of the males is to produce a large number of gametes. An individual that produces sperms does not require much energy to produce each gamete; therefore, it is capable of producing a large number of gametes. As in the case of the previously described hawk and dove game, if a majority of the individuals of the species are female, the males will be able to produce a large number of offspring. However, this does not imply that the females will disappear. When the number of males increases, a female, who produces ova, will be able to produce more offspring than a male. This game affects a balance in which the number of males and females is equal in most species.

Moreover, the two different strategies adopted by males and females must necessarily engender different secondary reproductive strategies. The results of these strategies are manifested as the sex differences in the appearance and behavior of males and females. Producing an ovum requires more energy than producing a sperm. Females spend more time mustering the energy required to produce ova. They are limited in terms of the number of ova they can produce and the number of offspring they can bear during their lifetime. For males, ova and females are limited resources. The demand for ova exceeds their supply. Males can produce far more sperms than females can produce ova. Males have the potential to mate frequently and produce a large number of offspring; however, there exists the possibility that they may produce few offspring. This is why males compete for females.

In certain species, the males that win this competition with other males can secure females and mate with them. The males of such species have evolved bodies that are larger than those of the females. For example, the large body of a male elephant seal is the result of this factor. This process is known as "male competition." In certain species, the males alone possess features much like weaponry, such as tusks and horns. Most of these weapons may be used not to protect females or infants from other animals but to compete with other males of the same species and secure the right to mate with females.

Females, on the other hand, are limited in terms of the number of offspring they are able to produce. Therefore, in order to ensure that their offspring are more likely to survive and in turn be capable of producing as many offspring as possible, it is necessary for females to carefully select appropriate males that are suitable for mating. This is known as "female choice." If a female selects a male with good genes, she can bear offspring who will inherit these genes. It may also be advantageous to females to select males who will bring them food, who will provide a good home range, or who will help them to raise their offspring. If females select males who provide good food or a good home range at present, they are simultaneously selecting the genes of males who have the ability to obtain good food and a good home range.

In certain species, however, the females select males who have beautiful physical features that are disadvantageous to their survival. For example, male peacocks have large and beautiful tail feathers. These features may indicate that they have good genes. It is unlikely that a physically weak male will possess beautiful features because of his bad genes. Beautiful males are able to survive despite their inconvenient ornamentations because they have good genes. This is why females are able to select males on the basis of their ornamentations. In addition, assume, for example, that the females initially preferred males who possessed weapons that gave them an advantage during male competition. Once a majority of the females were attracted to and selected such males, they were able to bear male offspring who, in turn, would be popular among the females. There may exist a runaway process by which male peacocks, in order to attract females, developed their beautiful characteristics to an extent to which those characteristics became disadvantageous to their own survival.

This mechanism, in which either of the sexes evolves specific characteristics, is known as “sexual selection.”

In the case of primates, individuals first select the physical, ecological, and social environments they live in. The choice of these individuals determines the type of the social group. Second, the type of the social group affects the reproductive strategies of the males and females in the social group. For example, in a one-male group, which consists of only one resident adult male and multiple females, the males will generally compete with other males by competing specifically for the position of being the only male in the group. In the species that form a one-male group, the males have evolved larger bodies than the females because a large body size is advantageous in fights.

On the other hand, in a multi-male, multi-female group, as formed by Tibetan monkeys, males have to compete not only to become members of the group but also to mate with as many females as possible within the group. Generally, males in a multi-male, multi-female group have dominance rank relationships and the subordinate males are prevented from mating in the presence of the dominant males. These dominance rank relationships serve to regulate the competition for females as well as other resources, such as food and space. However, this does not imply that low-ranking males cannot mate at all. Since they live in the woods, these males have the opportunity to mate without being caught by the dominant males.

Male competition may also occur after copulation; in this case, the sperms of one male compete with the sperms from other males in order to unite with an ovum. For example, male chimpanzees in a multi-male, multi-female group have large testicles relative to their body size, compared to male gorillas in a one-male group. Because of various competitions between males for females, the males are in a tense situation in a multi-male, multi-female group.

Further, most males belonging to the *Macaca* species emigrate from their natal group while the females do not. The females live in their natal group with relatives such as their mothers, sisters, and aunts, and when they give birth, they also live with their daughters, nieces, and grandchildren. On the other hand, most adult males who live in the same group are unrelated. They cannot disseminate their genes by assisting their relatives to mate more often with females; instead, they themselves mate and produce offspring. This holds true not only in the case of the competition for mating but also in the case of the competition for other resources such as food. For this reason, the potential tensions among males are always stronger than those among females although females may have the same kind of tensions.

The greater the number of sexually matured adult males in the group, the more intense will be the competition among them for the opportunity to mate with the females. The groups formed by Tibetan monkeys include several males.

Although I am unaware of the exact reason, ecological factors may have initially exerted an influence on the number of males. For example, if competition between groups for food resources is high, groups that include many males may have an advantage over other groups. Within a group, however, competition between males for females is high, if the ratio of males to females in the group is high. In the case of the Tibetan monkeys in the Yulingkeng group between 1985 and 1992, the average ratio of adult females to adult males was 1:0.92. There were almost as many adult males as females. Bonnet monkeys (*Macaca radiata*) share the common features with Tibetan monkeys. Bonnet monkeys also form a multi-male, multi-female group, and the ratio of adult males to adult females in this group is also high. They are similar to Tibetan monkeys in that they engage in friendly behavior that involves physical contact, and they are very tolerant of each other.[(1)] On the other hand, in Japanese monkeys or rhesus monkeys, the ratio of adult males to adult females in the group is low, and it is believed that the males do not interact with each other to a great extent. As observed in the case of Tibetan monkeys, the large number of males in the group require "lubricating oil," that is, friendly behavior involving physical contact such as penis sucking, in order to get along with one another. Conversely, it may be held that since male Tibetan monkeys engage in such social behavior, several males can live in one group. This is why male Tibetan monkeys frequently engage in friendly behavior that involves physical contact, and also display a rich variation in such behavior.

2. Bridging behavior between males

In addition to the behavior in which male Tibetan monkeys suck each other's penis in various ways, they display another interesting social behavior. This is not behavior in which males have other males suck their penises, but in which they display the penis of an infant to other males and suck it. In this type of behavior, two adult individuals lift the infant up together and suck or touch the penis of the infant. In this book, this behavior is referred to as "bridging behavior." The bridge in the phrase "bridging behavior" recalls a bridge spanning a river. This is because the infant that is held up resembles a bridge that is constructed between two individuals (Fig. 4–2) (Photo. 4–6). Bridging behavior is performed between two females or between a male and a female, but is most frequently performed between two males. First, I will introduce an example of bridging behavior.

Fig. 4-2 Bridging behavior

Photo. 4-6 Bridging behavior.

Observational case: An example of interactions involving bridging behavior

It was January 6, 1990, and it was a cloudy day. The Tibetan monkeys continued mating even after the arrival of the new year. At 9:53 a.m., the monkeys finished eating the corn and began climbing the cliff. They were on their way to the usual rock to rest there. Some monkeys were still eating corn at the provisioning site. These were the alpha male, Huangmao, the second-ranking male, Wushang, and the females, Zhi, Hua, and Tougou. The infants, Sanjiaolian, Keai, Zhiming, Guer, and Huagang, were playing around the provisioning site. Their mouths were full of corn. The juvenile male, Sanxiong, appeared undecided. He looked as though he was trying to decide whether he should follow the group members who had gone to the rock or should still look for corn stuck between the rocks.

Suddenly, at the provisioning site, Sanjiaolian screamed at Tougou. Since Sanjiaolian's mother, Baijamao, who would have protected her, was not present at the site, her behavior was very bullish. The alpha male, Huangmao, supported Sanjiaolian and chased Tougou. However, when he caught up with her, Huangmao did not punish Tougou by biting her but mated with her. Sanxiong approached the pair that was mating but was threatened by Huangmao. Sanxiong soon ran away and returned to his original place. The copulation ended after Huangmao ejaculated successfully.

At this juncture, Sanxiong approached Zhiming, Zhi's son, a male who was not yet one year old. Sanxiong held Zhiming and sat down with Zhiming facing his belly. I assumed that he would take Zhiming to Huangmao. However, he merely released Zhiming.

Zhiming soon returned to the spot near the provisioning site and began playing with the other infants. He jumped onto the face of the one-year-old female, Sanjiaolian, and placed his penis against her. This type of penis sucking was often conducted between two male infants, and sometimes between a male infant and a female infant. However, the ones who jumped on top of the others and had their genitalia licked were always male infants. Zhiming left the playing area again and wandered around. He approached Youshizhi. Youshizhi held Zhiming, but released him after some time.

The juvenile male, Duyanlong, held and groomed a one-year-old male, Guer, on the slope of the cliff. Huagang, a male who was not yet one year old, approached Sanxiong. Sanxiong exhibited his buttocks to Huagang and pointed to his buttocks with his hand, signaling, "Mount me." Huagang mounted Sanxiong. Guer, who was watching this, left Duyanlong's belly and approached Sanxiong and Huagang. Sanxiong was on the verge of taking hold of Guer, but stopped himself. Even if he did take hold of Guer, Huagang, who was beside him, was too young and small to be his partner in bridging. Sanxiong was not very inclined toward bringing with Huangmao, either. Duyanlong, who was left alone, also approached. Duyanlong, Sanxiong, Guer, and Huagang sat beside each other on a branch. Duyanlong groomed Sanxiong. The juvenile males and the male infants got along well. Bridging often took place when they assembled in this manner. However, Hua suddenly rushed to the site. She chased away Guer, who had been sitting near her son, Huagang. Guer ran away from her, as did Duyanlong and Sanxiong. The four males dispersed.

Nevertheless, both Duyanlong and Sanxiong moved to another tree branch and Duyanlong groomed Sanxiong again. Huagang was still with his mother, Hua, but Guer and Zhiming settled down near the branch of the tree on which Duyanlong and Sanxiong were seated.

At 10:06 p.m., Youshizhi, who had left the provisioning site, returned. As he approached Duyanlong and Sanxiong, he paused for a short time at the spot where Zhiming and Guer were seated. He took hold of Zhiming and placed him on his belly. When Youshizhi began walking, Zhiming clung to his belly. Youshizhi approached Duyanlong and Sanxiong, who were sitting beside and grooming each other on the branch. Youshizhi climbed the tree trunk, jumped over Sanxiong's head, stood between Duyanlong and Sanxiong, and faced Duyanlong. Sanxiong, who was being groomed by Duyanlong, was now one position away from his original seat. I assumed that Youshizhi would bridge with Duyanlong. Youshizhi turned and presented Zhiming, who was clinging to his belly, to Duyanlong. Duyanlong took hold of Zhiming's legs and lifted him up. Zhiming came to resemble a bridge between these two individuals, with his shoulders being held by Youshizhi, and his legs, by Duyanlong. This is bridging behavior. Duyanlong placed his head between Zhiming's legs and sucked his penis. Youshizhi took hold of Zhiming's back with one hand, and with his other hand, touched Zhiming's penis, which was being licked by Duyanlong. Sanxiong, who was behind Youshizhi, observed this scene over Youshizhi's shoulder. However, he could not participate in the bridging. The bridging ended within three seconds. When Duyanlong released him, Zhiming clung to Youshizhi's belly again. I surmised that social grooming would occur between any two individuals from among Youshizhi, Duyanlong, and Zhiming. However, at that point, they were merely seated side by side. Youshizhi and Duyanlong, who had held Zhiming, were seated face-to-face on the branch. At this point, Zhiming slipped away from Youshizhi's belly. However, Youshizhi followed him in order to take hold of him and brought him back to Duyanlong. Duyanlong and Youshizhi performed the bridging behavior with Zhiming again. After this bridging, Zhiming clung to Duyanlong's belly. Finally, Youshizhi left Duyanlong and Zhiming and moved to the top of the cliff.

Next, Yanbaiban carried Huagang to Duyanlong, who was holding Zhiming, and to Sanxiong, who was beside him. Yanbaiban was a new immigrant adult male in this group. He sat next to Duyanlong but Duyanlong did not perform bridging. So, Yanbaiban decided to take Huagang to Wushang. He sat in front of Wushang and presented Huagang to him. However, Wushang did not bridge with him either, but soon stood up and left Yanbaiban. Yanbaiban released Huagang and followed Wushang.

Sanxiong returned to the provisioning site, where no adult male was present. Duyanlong was still holding Zhiming on the branch and grooming him. Two-year-old Chawei approached, and Duyanlong groomed him as well. Duyanlong, Zhiming, and Chawei later went up to the cliff together and disappeared from my sight.

This is an example of a series of interactions among Tibetan monkeys involving bridging behavior.

Behavior similar to the bridging behavior of the Tibetan monkeys is reported among the Barbary monkeys. Barbary monkeys belong to the same genus *Macaca* as the Tibetan monkeys. I have introduced them as the only macaque species distributed in Africa, while all the other macaques are currently distributed in Asia. Male Barbary monkeys also bring an infant to other males, and they lift the infant up together. This behavior of the Barbary monkeys was first reported by J. M. Deag and J. H. Crook in 1971.[(2)] In their paper, they termed this behavior "agonistic buffering" because they believed that its function is to "decrease the risk of attack that the individual may encounter during interactions with other males, by using an infant as a buffer." However, once a certain behavior is named on the basis its function, the name may later have to be changed when people begin to question whether the behavior really performs the function suggested by its name. In fact, D. M. Taub claimed that this behavior has another function, which I will discuss later. Further, Taub also proposed the use of a more neutral term for this behavior, "triadic male-infant interaction."[(3)] I realize that the use of this neutral term will not pose any problems. However, among Tibetan monkeys, at least, the behavior in which two individuals lift up an infant takes place not only between two males but also between a male and a female, and between two females. Therefore, the name of this behavior has to be changed to "triadic male-female-infant interaction" or "triadic female-infant interaction," which is inconvenient. In addition, names such as "agonistic buffering" and "triadic male-infant interaction" have come to be used to refer to behavior in which an individual holds an infant in front of another individual; this has been reported in baboons.[(4) (5)] In order to avoid any confusion with such behavior, I would like to suggest another name for the behavior in which two individuals lift up an infant together. This name is based on the nature of the behavior alone. In his 1977 article, A. Estrada used the term "bridging" in relation to a picture of two stump-tailed monkeys lifting up an infant together.(6) Thus, I use the term "bridging" to refer to the behavior in which two individuals lift up an infant together. While there have not been many detailed studies of this behavior, it has been reported in five macaque species: Assamese monkeys, crab-eating (long-tailed) monkeys (*Macaca fascicularis*), Tibetan monkeys, Barbary monkeys, and stump-tailed monkeys. Although we use the term "bridging" to refer to "behavior in which two individuals lift up an infant together," bridging among Tibetan monkeys and among Barbary monkeys, for example, differs in various respects. Yet, there exist several similarities as well. On the other hand, I would like to draw attention to the fact that social interactions among baboons are quite different from bridging behavior. Bridging behavior may have evolved in the macaque species alone.

3. The function of bridging: The agonistic buffering hypothesis

Primates rarely display behavior such as giving or handing something to another individual. There are a few cases in which chimpanzees that are begged for food, such as meat, will hand over the food to another chimpanzee. In bridging behavior, although an infant is presented to another individual, this does not imply that the monkeys intend to give the infant to the other individual. However, since they make the effort to carry the infant and hand it to the other individual even without being asked to do so, this behavior is very interesting. What could possibly be the function of bridging behavior?

When I completed my observations spanning the period from the winter of 1989 to the spring of 1990, I had the following impression of bridging behavior among male Tibetan monkeys.

As opposed to the behavior of baboons, who hold an infant in front of another male when they are about to be attacked or when in a tense social situation, bridging behavior among Tibetan monkeys often occurs when the group members are relaxing.

A few infants often play near the group members when most of them are resting. On such occasions, a male approaches the playing infants and takes hold of one of them. An infant sometimes approaches and is held by a male. The male holds the infant for a while, ranging from a few seconds to more than 10 minutes. Since the males sometimes release the infant quickly, the behavior that involves holding infants does not necessarily lead to bridging. However, the male who holds an infant often takes it to another individual. Sometimes, the male carries the infant to one of the group members beside him, to an individual who is at a distance, and occasionally to an individual who is even more than 50 m away. When two males face each other, trapping the infant between them, they display a distinctive facial expression, stretching their lips to the sides of their faces, moving them, and gnashing their teeth. One male holds the thighs or buttocks of the infant while the other holds its shoulders, and together, they turn the infant on its back. The two males then lift up the infant together for a few seconds. This is bridging behavior. While lifting up the infant, either or both of the males often lick the infant's genitalia or touch it with their fingertips. Other individuals occasionally join the two adults and the infant engaged in the bridging behavior, and three or more males may lift up the infant together. The actions that follow the bridging behavior vary. Sometimes, the three individuals disperse quickly; however, the two males often groom each other with the infant between them, or groom the infant. Occasionally, either male takes hold of the infant again and carries it to another individual. During the social interactions that involve bridging behavior, the infant does not resist the adult males and allows them to do what they like. The mothers, too, tolerate other individuals' using their offspring in bridging.

In order to explore the implications of bridging behavior, I recorded the series of events that preceded and followed this behavior during my study periods from the winter of 1990 to the spring of 1991. During these periods, I observed a total of 333 cases of bridging behavior. After returning to Japan, I transferred the records of the behaviors from my fieldnotes to a computer in succession and analyzed the data.

First, I examined the social contexts in which bridging behavior occurs. On an average, each male engaged in bridging behavior 0.35 times per hour. On summarizing this data, I discovered that there was virtually no agonistic interaction in the group immediately preceding bridging behavior. Nevertheless, 35 (10.5%) bridgings occurred after a confrontation in the group. However, on further categorization of the contexts, I identified only four cases (1.2%) in which a male who had been attacked took hold of an infant in order to perform bridging and carried it to the male who had attacked him. The instances of bridging that occurred in agonistic interactions constituted a very small number. On the contrary, the male Tibetan monkeys engaged in mounting immediately after agonistic interactions. I am of the opinion that the males frequently engaged in mounting immediately after an agonistic interaction and after the mounting, the two males appeared to have entirely forgotten their confrontation. In 18 cases (5.4%), the bridging behavior occurred between a male who was involved in a confrontation and a male who was not. In 13 cases (3.9%), the bridging behavior occurred between two males who were not involved in fights even in the case of a confrontation in the group. In the other 298 cases (89.5%), the bridging was not preceded by any agonistic interaction in the group. Thus, the cases in which bridging occurred when the social tension was high due to agonistic interactions were rare.

However, it was observed that at the provisioning site, where all the members of the group created a tense and exciting atmosphere immediately before the scattering of the corn, the monkeys frequently engaged in a variety of social behaviors, including bridging and mounting. During the winter, there were few substantial food resources in the woods. It was particularly amusing to observe the behavior of the monkeys that were waiting for corn at the provisioning site. However, since such a situation is artificial, I eliminated it from my data.

Second, I analyzed the events that occurred in the 827 cases in which two males encountered each other in the woods. I will now enumerate the occasions on which an infant was involved in such an encounter between two males. Immediately before bridging, the two males face each other and one of them holds the infant. I categorized the manner in which this situation occurs into three different patterns. The first pattern is one in which a male takes hold of an infant and carries it to another male. The second is where a male approaches another male holding an infant. The third is the pattern in which one of two males who are already in close proximity to each other takes hold of an infant. Bridging can occur immediately after one of these three patterns.

The first among these three patterns provides the clearest indication regarding which of the two males wishes to initiate the bridging. The individual who approaches holding an infant to his belly must want to bridge with the other. The male who carries the infant presents it to the other male. If the individual who is presented with the infant is inclined toward bridging, the two males perform the bridging behavior. In actuality, the bridging occurred in 185 cases (58.5%) from among the 316 cases in which one male carried an infant to another.

In the case of the second pattern, it is difficult to judge whether the individual holding the infant is attempting to bridge with another male or merely holding the infant. How is the individual who approaches another male holding the infant? This individual may approach another male for bridging or may approach another male who is incidentally holding an infant. It is impossible to distinguish between these two cases. Therefore, I analyzed all these cases together, and observed that bridging occurred in 70 cases (60.9%) from among 115 instances of the second pattern.

The last is the case in instances in which a male who is not holding an infant approaches another male who is not holding an infant. In 51 (12.9%) of the 396 encounters in which neither of the males had an infant, one of the males took hold of an infant near him. In these cases, it may be assumed that the male who took hold of the infant attempted to bridge. However, I received the impression that this third pattern indicated a situation in which both the males wanted to bridge, and they took hold of the infant for bridging almost simultaneously. In fact, in the third pattern, bridging occurred in 43 (84.3%) of the 51 cases in which one of the males took hold of an infant near him; this accounts for a considerably large proportion.

Encounters between males involving an infant resulted in a comparatively high ratio of bridging. However, there may have been certain cases in which the males wished to bridge but there was no infant in the vicinity. I was unable to divide my data on the basis of the situations in which the males did not engage in bridging despite the presence of infants, and the situations in which they could not engage in bridging due to the absence of infants in the vicinity. Since Tibetan monkeys live in densely populated groups, it was extremely difficult to record the data regarding which three individuals were present within how many meters of each other.

Nevertheless, it is certain that greeting behavior often occurred between two males in cases in which neither of them was holding an infant during their encounter. As I have already mentioned, the males engaged in presenting and penis showing during their encounters. The male who approached another performed the presenting behavior in 26 (7.5%) of 345 cases and performed the penis-showing behavior in eight cases (2.3%). Except for one instance, the behaviors in all these cases were directed by subordinate males toward dominant males. Mounting occurred in 43 cases (12.5%), embracing in 17 cases (4.9%), and penis sucking in three cases (0.6%). In contrast, only one instance of mounting (0.2 %) and three of embracing (0.6 %) occurred in the 428 encounters in which either male was holding an infant. These represent the extremely rare cases in which one of the two males holding an infant engaged in mounting or embracing. Thus, in encounters between two males, bridging frequently occurs when either of the males holds an infant, while friendly behavior that involves physical contact such as embracing and mounting frequently occurs when neither of the males holds an infant.

Next, I examined the actions performed by the males after bridging. It is important to examine the proximate functions of bridging behavior. On summarizing my data, I discovered that after the bridging behavior, the males frequently sat down together and engaged in grooming. It was infrequent for the males to sit down together and engaged in grooming after the bridging behavior did not occur, because the male who was presented with an infant rejected the bridging or because the male did not present an infant upon being approached by another male. Compared to these cases, the two males more frequently groomed

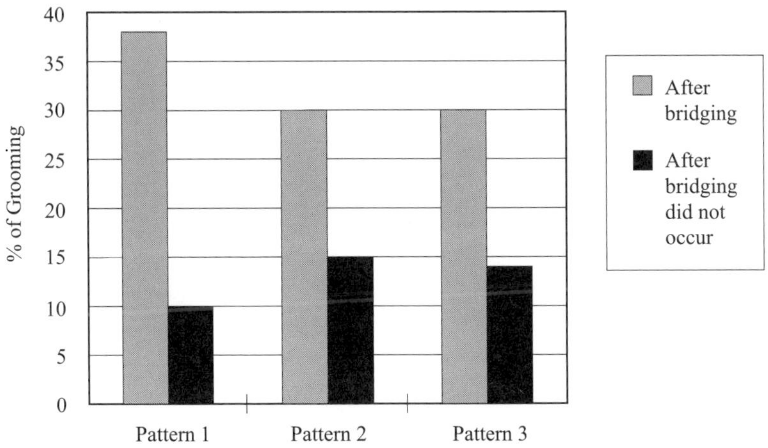

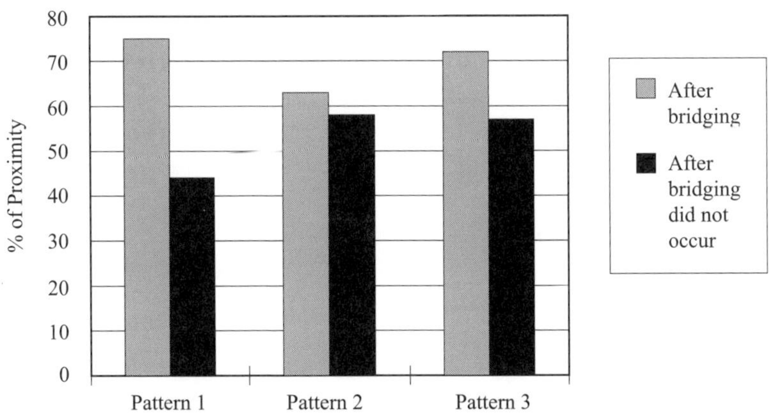

Fig. 4–3 Behavior engaged in by the males after bridging.
The percentage of social grooming between males and close proximity (being within reach) for more than 1 min, after bridging did and did not occur. After bridging: Bridging occurred when one of the males presented an infant to the other. After bridging did not occur: Bridging did not occur because a male did not receive an infant presented to him by another male. Pattern 1: A male who was holding an infant approached another male. Pattern 2: A male approached another male who was holding an infant. Pattern 3: One of the two males who encountered each other held an infant.

each other and sat together after the bridging (Fig. 4–3).

Further, there were nine cases (7.4 %) in which the males were rejected and attacked even after they had carried and presented an infant to another male. There were two cases (4.4%) in which the males who approached another male with an infant were attacked by the male who was approached, without the bridging being performed. In contrast, no aggressive interaction occurred after the bridging behavior.

Finally, I examined which individuals attempted to bridge frequently and which individuals succeeded in bridging.

During my study period, the juvenile males, Duyanlong and Sanxiong, were the lowest-ranking males, excluding the infants. It was these juvenile males who frequently approached other males in order to perform bridging. Their bridging partners were often higher-ranking males such as Huangmao and Wushang. Sanxiong often carried an infant to the juvenile male, Duyanlong, who was like an older brother to him, too. Duyanlong was exclusively presented with an infant by Sanxiong, who was like a younger brother to him. However, when Duyanlong became the alpha male the following year, the other males, too, presented infants to him.

Based on these facts alone, this tendency may be determined by each individual's personality and preference. Therefore, I analyzed the relation between the dominance ranks of the males and the frequency of the bridging behavior. Consequently, I discovered that lower-ranking males frequently attempted to bridge with higher-ranking males, shown in detail as follows.

First, the higher the rank of the individual, the more often he was chosen as a partner for bridging. There was a statistically significant correlation between the dominance ranks of the males and the frequency with which they were approached by other males with infants. Moreover, the lower the rank of the individual, the more often he carried an infant to another male although this was not a statistically significant correlation. With regard to the dominant-subordinate relationships in each pair of any two males, I found a statistically significant difference that the frequency with which the subordinate male carried an infant to the dominant male was higher than the frequency with which the dominant male carried an infant to the subordinate male. However, the encounters between two males were generally initiated by a subordinate male who approached a dominant male. On the basis of this analysis alone, the tendency of the subordinate males to frequently carry an infant to the dominant males may be a result of the simple fact that they approached the dominant males more frequently. Therefore, I examined how often a subordinate male carried an infant when he approached another male. A subordinate male was more likely to hold an infant when he approached a dominant male than vice-versa (Fig. 4–4).

In addition, I examined the frequency with which a male who was holding an infant was approached by another male. A correlation revealed that when the males were holding an infant, the higher-ranking males were more likely to be approached by another male.

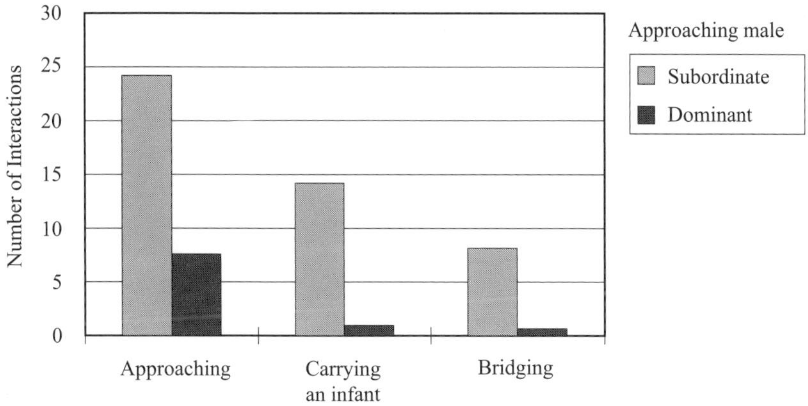

Fig. 4-4 Influence of dominance ranks on the encounters between the males.
Approaching: All the cases in which one male approached another. Carrying an infant: A male who was holding an infant approached another male. Bridging: Bridging occurred when a male carried an infant to another male.

In summary, among male Tibetan monkeys, the lower-ranking males are more likely to perform bridging with other males, and even the high-ranking males are more likely to bridge with the dominant males.

The frequency with which bridging occurred was also associated with the familiarity between the two males, as well as their ranks and their dominant-subordinate relationships. Bridging frequently occurred between two familiar males. In other words, the males in the pairs that frequently engaged in bridging were familiar with each other. As in the case of the analyses of the social relationships among primates, I used the frequency of grooming as the index of familiarity. The more frequently two males performed bridging, the more frequently they groomed each other.

Let us consider this aspect in greater detail. The males often groomed those males who frequently carried an infant for bridging. However, this may be a consequence of the fact that subordinate males frequently groomed dominant males and initiated bridging more frequently. An unexpected fact, however, was that at least during the period of my study, the frequency with which subordinate males groomed dominant males and the frequency with which dominant males groomed subordinate males did not display a statistically significant difference. In addition, in cases in which a male A frequently approached another male B in order to perform bridging, male B, too, often approached male A for bridging. Further, the males often carried an infant in order to bridge with a male who groomed them frequently although this was not a statistically significant difference. Finally, the males frequently groomed those who groomed them in return. In summary, the males who shared close relationships in which they frequently groomed each other also engaged in bridging with each other frequently. This result is consistent with the fact that males often groom each other after bridging.

On the basis of these results, we will now consider the function of bridging behavior in macaque species.

The meaning or the function of the behavior in which a male monkey holds or carries an infant to another male has been a subject of controversy. In 1959, Jun'ichiro Itani observed that low-ranking male Japanese monkeys entered the central area of the group—composed of high-ranking males and females—carrying an infant, and wrote that low-ranking males are using the infant as a passport to enter the central area of the group.(7) This was the first proposition suggesting that monkeys use an infant to manipulate their relationships with others. Later, on the basis of a series of interactions among Barbary monkeys in which a male carried an infant to another male and they both lifted the infant, Deag and Crook proposed the agonistic buffering hypothesis, which suggested that the function of bridging behavior is that "low-ranking males reduce the possibility of being attacked by high-ranking males." Subsequently, as already discussed, the term and the concept of "agonistic buffering" were also applied to the behavior of other primates, and in particular, to the interactions of baboons in which one individual holds an infant in front of another individual. In actuality, male Bonnet monkeys, for example, are less likely to be attacked by high-ranking males while they are holding an infant as compared with when they are not holding an infant.(8) My observations may also indicate that male Tibetan monkeys are less likely to be attacked by other males while they are holding an infant. However, this is difficult to evaluate accurately. The monkeys may be holding an infant cautiously because they are about to be attacked, or the males holding an infant may approach another male when they know that they are not going to be attacked. Nevertheless, the behavior in which male Tibetan monkeys hold and carry an infant to other males was more frequently directed by subordinate males toward dominant males. In addition, the male Tibetan monkeys carried an infant more frequently when they approached the dominant males. On the basis of this tendency, it may be assumed that the behavior in which subordinate males carry and present an infant to the dominant males functions as a form of appeasement and is used to greet and gratify other males. That is, infant presentation, like presenting and penis showing, is an appeasement behavior. In fact, subordinate males who approach dominant males without an infant sometimes perform presenting or penis-showing behaviors. Thus, if we consider this behavior in light of why the males present an infant instead of their mere actions or expressions, the agonistic buffering hypothesis will hold true. When subordinate males approach the dominant males while holding an infant, they reduce the possibility of attacks by using the infant as a buffer since they run the risk of being attacked if they do not carry an infant.

However, Tibetan monkeys that approach other males not only hold an infant but also lift the infant and lick his penis along with the other male. I will now examine this behavior. Bridging occasionally occurs when the tension within the group is high on account of an agonistic interaction. In this case, bridging may mitigate the heightened tension between the males, in a manner similar to penis sucking and embracing in the same context. However, unlike the social interaction among baboons, bridging among Tibetan and Barbary monkeys rarely occurs in tense situations, for example, immediately before or after an agonistic interaction. A male holding an infant occasionally makes the effort to carry and present the infant to a male who is at a distance of more than 50 m. These males have not engaged in any interaction with each other immediately before this infant-carrying behavior. Therefore, bridging is not necessarily performed in order to avoid the direct aggressive behavior that occurs as a result of an agonistic relationship or to resolve the tension that already exists. Further, dominant individuals sometimes carry an infant to subordinate individuals in order to perform bridging, although the frequency of this occurrence is lower than that of the reverse case. Since the approaching dominant individual runs no risk of being attacked by the other male even when he does not carry an infant, it is difficult to apply the agonistic buffering hypothesis to this case. Moreover, baboons or Japanese monkeys who are approached by a male carrying an infant do not receive the infant and lift it along with the approaching male. These monkeys merely hold the infant in front of another male. On the other hand, in some species belonging to the genus *Macaca*, such as Tibetan and Barbary monkeys, the bridging behavior acquires a ritualized form in which two males lift an infant together.

Among the variety of friendly behaviors displayed by male Tibetan monkeys, bridging, like penis sucking, is a species-specific behavior in which the monkeys lick the penis of the infant. The form of the behavior in which the males display and have their penis licked by another male is similar to the form of the behavior in which males present an infant to another male and have its penis licked. Male Tibetan monkeys are preoccupied with the existence of the penis during bridging. In fact, male infants are used in bridging between males more frequently than female infants. When a male presents an infant for bridging, the behavior may be more effective if the infant has a penis. If this hypothesis is true, the monkeys, instead of displaying their own penis and inviting penis sucking, may be using the penis of the infant as a substitute. The presenting of infants, which functions as an appeasement behavior, is probably more effective than penis showing. When I tried to present this hypothesis in a scientific journal, one of the referees stated, "Eliminate this sentence because it is too Freudian." Therefore, I presented this hypothesis in my own thesis, that is, during bridging behavior, a monkey may be using the penis of an infant as a substitute for his own penis because the use of the infant's penis is more effective.

After a male who had been presented with an infant received the infant and performed the bridging, the two males often sat together and groomed each other. This indicates that bridging facilitates the social grooming that follows it. The grooming and close proximity engaged in after bridging account for a large proportion of the total grooming and close proximity between the males. Bridging often occurs between two individuals who share a close relationship, as is indicated by their frequent grooming. Further, during bridging, males use the infant or the infant's penis as a substitute for their own penis and assume a posture similar to that assumed during penis sucking. As presented above, on considering the attributes of the individual who performs the behavior, the combinations of the individuals, the form of the behavior, and the context in which it is performed, bridging can be understood as a type of friendly behavior of which function is to reduce potential tensions and to establish and maintain the social bond among the males.

We can now interpret the bridging behavior between male Tibetan monkeys as a behavior in which the males use an infant as a social tool in order to manipulate their social relationships with other males. It is a well-known fact that chimpanzees use tools. They obtain their food by using tools in various ways, such as fishing ants or termites with a twig or cracking nuts with a stone. The goal of the chimpanzees is to obtain food, and their tools are objects such as twigs or stones. In contrast, it is interesting to note that in the case of Tibetan monkeys, the goal of bridging is to manipulate the relationships with others and the tool is an infant monkey of the same species.

Among Tibetan monkeys, the infants function as a bridge that enables the large number of males in the group to cope with one another. The Japanese proverb says "A child is the bond between husband and wife." This phrase can be refashioned as "A child is the pledge of the group" for Tibetan monkeys.

4. Males have their favorite infants: The baby-sitting reinforcement hypothesis

Thus far, we have examined bridging solely in terms of the two adult males among the three individuals who are involved in this behavior. However, the identity of the third individual that participates in the bridging – the infant – is also significant to the male Tibetan monkeys. Male Tibetan monkeys select the infant accordingly.

I have already mentioned that male infants were more likely to be used for bridging than female infants. Each male infant born during the spring of the year was used for bridging 0.43 times per hour, while each female infant was used only 0.04 times per hour. Infants that were three years old or younger were used for bridging. Among these, one-year-old infants were used more frequently than two-year-olds, and infants that were not yet a year old were used more frequently than one-year-olds. This implies that males that are not yet a year old possess characteristics that make them the most likely to be used for bridging; this is, in fact, true. Male Tibetan monkeys may well be aware that bridging is more effective if they use younger, smaller, male infants.

The infants used for bridging were those that were frequently held, carried, and groomed by adult males. Males were more likely to be held than females in the same age, one-year-olds were more likely to be held than two-year-olds of the same sex, and zero-year-old infants were more likely to be held than one-year-olds.

However, the factors affecting the frequency with which the infants were used for bridging and the factors affecting the frequency of their interactions with adult males, such as being held or groomed, were not determined solely by characteristics such as the sex, age, or body size of the infants. It is interesting to note that most of the male Tibetan monkeys frequently held one specific infant from among all the infants in the group and used that infant for bridging. Since there were several infants of the same age and the same sex, I could not identify them unless they were marked with hair dye. However, once I managed to identify them, I observed that each adult male frequently held one infant in particular. For example, during my 88-day observations—that is, from December 5, 1989 to January 21, 1990 during the winter, and from March 12 to April 20, 1990 during the spring—Youshizhi frequently held Zhiming; Duyanlong, Huagang; and Sanxiong, Zhiming. Table 4–1 presents the number of infants held by each male during these periods. I will refer to such an infant as the male's "favorite infant," and to the male as the infant's "familiar male." A familiar male and his favorite infant often groomed each other and established an extremely close relationship. At night, the favorite infant sometimes slept on his familiar male's belly. Since the infants groomed the male while they were being held, it was unclear whether the infants also had active interactions with their familiar males. However, I observed several instances in which it appeared that the males who had rarely held an infant could not take hold of him easily, or when the infant, even while playing with the others, jumped to his familiar male to be held when the male signaled to him by smacking his lips. In any case, the infants

Table 4-1 Frequency with which the males held infants.

Male	Infant Name / Age / Sex	Zhiming 0 Male	Huagang 0 Male	Baixiang 0 Female	Guer 1 Male	Sanjiaolian 1 Female	Keai 1 Female	Total
Huangmao		5	5	10	2	0	0	22
Wushang		15	7	6	2	0	0	30
Erxiong		6	8	26	2	0	0	42
Youshizhi		105	13	4	16	0	0	138
Chiyan		1	0	0	1	0	0	2
Yanbaiban		9	20	0	2	0	0	31
Duyanlong		30	74	1	16	0	0	121
Sanxiong		68	36	2	17	0	0	123
Total		239	163	49	58	0	0	509

The table presents the number of cases in which males held infants during 88 observational days between Dec. 5, 1989 and Jan. 21, 1990 and between March 12, 1990 and April 20, 1990. The males who were 4 years of age and older and the infants who were 0–1 years of age are listed. The males are arranged in the descending order of their dominance ranks.

recognized their own familiar males.

Why do male Tibetan monkeys take hold of a certain infant more often than they do other infants? D. M. Taub observed a similar relationship among Barbary monkeys and surmised that the favorite infant may be kin-related to his familiar male through a matrilineal line, that is, the infant and his familiar male may be brothers who have the same mother, or the infant may be the son of the male's sister, who has the same mother as the male. The males would wish to take good care of their brothers or nephews, who are their maternal kin, and assist their successful survival. According to the theory of kin selection, this enables the monkeys to ensure the survival of a large number of individuals who share common genes with them and would increase the inclusive fitness. Thus, males diligently look after their kin, such as their maternal brothers. In addition, D. M. Taub proposed the reinforced baby-sitting hypothesis.(9) He assumed that males may not only look after their kin themselves, but also, for example, carry their brothers to their uncles and indicate, "This infant, like me, is also your nephew. Please take care of my young brother, too." This activity of announcement may be expressed as bridging behavior in which the male presents his brother to his uncle. When the brother reaches adulthood, he carries his nephew to the older brother who has looked after him so far, and announces his nephew by using him for bridging with the older brother. Thus, the male appears to say, "Since I have a nephew, let us now take good care of him, Brother." In summary, maternally related males reinforce the care provided to the infant by identifying the infants that are their kin through bridging. If true, this is a very interesting hypothesis. Unfortunately, however, since the study conducted by Taub could not reveal the kin relationships within the group, it was unable to confirm that related male Barbary monkeys actually engage in such behavior.

5. Presenting a favorite infant to the recipient male: My friend's friend is my friend

Did the Tibetan monkeys at Huangshan perform bridging in accordance with Taub's hypothesis? Duyanlong and Sanxiong, who were born in this group, were the individuals that frequently engaged in bridging and held infants. Duyanlong often held Huagang and Sanxiong often held Zhiming. However, Duyanlong was not Huagang's brother, and Sanxiong was not Zhiming's brother. If the reinforced baby-sitting hypothesis were correct, Duyanlong ought to have held and used his sister, Baixiang, for bridging. On summarizing the data obtained during the period from 1989 to 1992, I was unable to obtain a positive result that Tibetan monkeys held and used their maternal siblings and their maternal nephews or nieces born from a common mother as the infants for bridging. Such factors did not influence the males' choice of the infant they made their favorite. Then, how did they determine which infant was their favorite?

Certain observations provided a clue to the solution of this problem. One of these is that after an infant was used for bridging, the male who was presented with the infant received and took the infant to another male, and the infant was sometimes passed around among multiple males in a manner resembling a relay. Similar events sometimes occurred while the infant monkeys were playing together. If one of the infants incidentally began playing with a certain object, for example, an empty can, all the others concentrated their attention on that particular object and were preoccupied with securing that object in order to play with it, despite there being other objects in the vicinity.

I also observed that Baixiang, a female who was not yet one year old and who had not often been used for bridging, was extensively used by multiple males, especially the high-ranking males Huangmao, Wushang, and Erxiong, over one week during January 1990 (Fig. 4–5). Baixiang exhibited no unusual changes during this period. Undoubtedly, she did not suddenly turn into a male, nor did she approach adult males more frequently than usual. I surmised, "The very fact that certain males used Baixiang for bridging led to her being used extensively for bridging." If certain males favor a particular infant for certain reasons, do the other males also begin to use the infant for bridging? After returning to Japan in the spring of 1990, I analyzed the data I had gathered up to this point, and examined which male carried an infant to another male. I

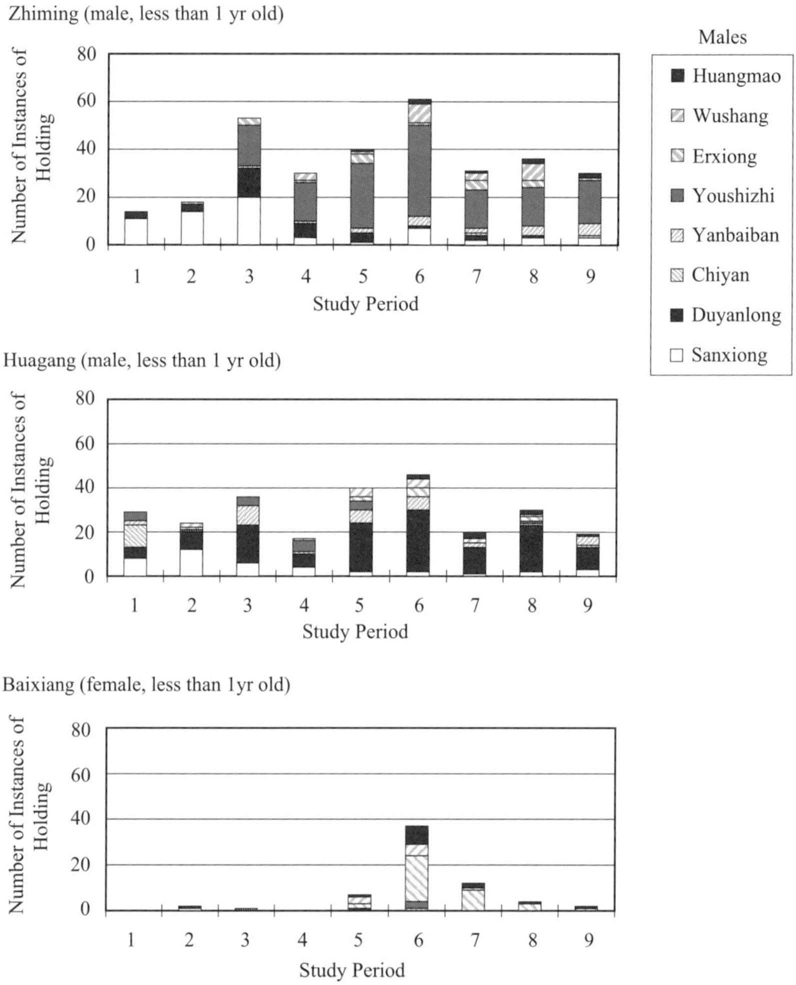

Fig. 4–5 Number of instances in which infants were held in each study period.
Study period: 1 = Dec. 13, 1989–Dec. 21, 1989; 2 = Dec. 22, 1989–Dec. 31, 1989; 3 = Jan. 1, 1990–Jan. 10, 1990; 4 = Jan. 11, 1990–Jan. 19, 1990; 5 = March 12, 1990–March 16, 1990; 6 = March 17, 1990–March 25, 1990; 7 = March 26, 1990–April 3, 1990; 8 = April 4, 1990–April 11, 1990; 9 = April 13, 1990–April 20, 1990.

discovered that male Tibetan monkeys chose an infant for bridging depending on the identity of their partner in bridging. They carried an infant to its familiar male. In other words, the males were more likely to be presented with their favorite infant (Fig. 4–6).

During my observational period, I recorded 713 instances of infants being

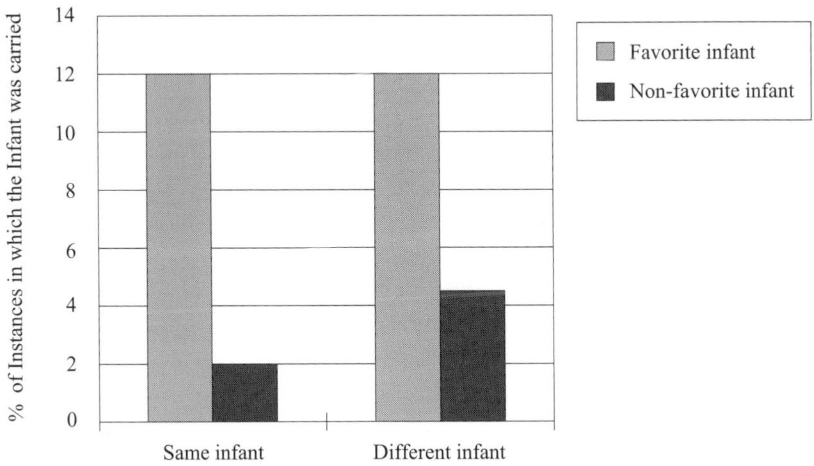

Fig. 4–6 Favorite and non-favorite infants used in bridging behavior.
The percentage of all the cases in which an infant was held and either a favorite or a non-favorite infant was carried to a male. Favorite infant: A male was presented with his favorite infant. Non-favorite infant: A male was presented with an infant who was not his favorite. Same infant: The initiator and the recipient male had the same favorite infant. Different infant: The initiator male's favorite infant was different from that of the recipient male.

held by males and 240 cases of bridging behavior. Table 4–1 presents the number of cases in which the infants were held by the males. However, from among all these instances, I excluded the cases in which the males used the infant for bridging after holding it. This is because it was not possible to discern whether a male held an infant for bridging or merely for the purpose of holding it. The inclusion of the cases in which a male held an infant in order to bridge with a male who was frequently his partner in bridging leads to complications in the evaluation of the males' preferences for infants on the basis of the holding behavior. Therefore, I excluded from the data in Table 4–1 the cases of holding in which the male carried an infant to another male for the purpose of bridging. Thus, the number of instances of holding presented in Table 4–1 indicates the male's preference for the infant, which is in conjunction with the frequency with which he groomed the infant. I define a favorite infant as one that is most frequently held by a particular male. Since one of the new immigrant males, Chiyan, held infants only twice during this period and there was no statistically significant difference in how frequently he held them, I consider that he did not have a favorite infant.

Not only did the males frequently hold their favorite infant—this is implied by the definition itself—but they also frequently groomed that infant and used it for bridging.

In the cases in which the male holding an infant approached another male, the first of the three patterns of bridging behavior provides the clearest indication of which infant the male who initiated the behavior wanted to use for bridging with the other male. During my observational period, the males carried and presented an infant to another male a total of 229 times. I compared the frequency with which the males actually carried an infant "a" to its familiar male "A" when they held the infant "a," and the frequency with which the males carried infants other than "a" to the male "A" when they held infants other than "a." When the males held a certain male's favorite infant, they carried it to that male more frequently than when they held other infants (Fig. 4–6).[(10)]

This analysis is based on a comparison between the favorite and non-favorite infants of each pair of males. Therefore, the result of this analysis is not influenced by the tendency of certain males to frequently engage in bridging. The result implies that male Tibetan monkeys select an infant for bridging according to the relationship between the males who are their partners and the infant.

Taub observed that male Barbary monkeys who share the same favorite infant often engage in bridging behavior.[(9)] Taub assumed that "an infant and the two males who have a close relationship with the infant may be maternal kin," and proposed the reinforced baby-sitting hypothesis on the basis of this assumption. Among Tibetan monkeys, however, the two males who favored the same infant were not related. Is this tendency the result of the fact that two males who have the same favorite infant frequently bridge or that for certain reasons, two males who frequently bridge have the same favorite infant?

In the previous analysis, I did not differentiate between the pairs of males in which both favored the same infant and the pairs in which the two males favored different infants. Therefore, I conducted the same analysis by separating the abovmentioned pairs. I then discovered that even in cases in which the two males favored different infants, male Tibetan monkeys carried their partners' favorite infants to them more frequently than they did other infants.[(10)] In other words, when a male chose an infant for bridging, he ensured that the infant was not his own favorite but his partner's favorite.

This finding makes me speculate whether the male who is presented with an infant is happier when he receives his own favorite rather than any other infants. If a male is aware of which male favors which infant, he can use this relationship to his advantage. One aspect of bridging behavior is that by carrying an infant to a male who is close to that infant, the males attempt to form and maintain friendly social relationships with each other. In summary, bridging is a behavior in which the males attempt to get along with each other by carrying their partner's favorite infant and lifting it together. I propose a hypothesis that I refer to as the "My friend's friend is my friend" hypothesis. Male Tibetan monkeys do not merely use the infants for bridging but have also formulated a clever way of selecting the infants on the basis of the relationship between their male partner and the infants.

6. Recognition of the social relationships between other monkeys

Male Tibetan monkeys select infants for bridging based on the relationship between the infant and the recipient male. This indicates that monkeys are aware of the social relationships among other individuals, such as who is familiar to whom. In this context, bridging becomes an interesting approach by which to examine the monkeys' recognition of these relationships.

The members of several primate species recognize each individual who lives in their group. For example, human beings can identify a close friend by merely hearing his or her voice over the telephone. Monkeys, too, are able to identify individuals without actually seeing them merely by hearing their voices. If this ability is exploited, it will be possible to conduct playback experiments in which monkeys are allowed to listen to various recorded voices of the members of the group and their responses to these can be examined.

R. M. Seyfarth and D. L. Cheney examined the responses of wild Vervet monkeys (*Cercopithecus aethiops*) in a series of experiments and investigated how the monkeys recognize the world around them, especially their own social group.[(11)] First, they recorded the screams of a certain infant in the group, and after some time, they played back the recorded screams to other monkeys using a tape recorder hidden in the bushes. At this juncture, the mother of the infant exhibited a more conspicuous response to the scream than the other female monkeys who had babies. She looked toward the source of the screams and proceeded into the bushes to investigate. This experiment shows that mothers recognize the screams of their offspring and understand the relationship between their babies and themselves.

In addition, an interesting event occurred during this experiment. The female monkeys who had heard the scream that was not of their own offspring looked toward the source of the scream and then at the mother of the screaming infant, as if to say, "Your offspring has just cried. Is it okay to ignore your infant's scream, Mother?" This shows that the females knew who the mother of the infant was. That is, not only do primates recognize each individual in the group and understand the relationship between others and themselves, but they also understand the relationship between two other individuals, such as the relationship between an infant and its mother.

How do they acquire the knowledge of their social relationships and how do they use this knowledge?

To date, several studies have shown that primates utilize their understanding of the social relationships between others, especially to form coalitions in the case of agonistic interactions. When they form a coalition in an agonistic interaction, they use their knowledge of the dominance ranks of others. The understanding of the linear dominance relationships among individuals is important in agonistic interactions during which they compete for resources such as food and mates or for a higher rank in the group, which results in greater access to the resources.

Primates modify their behavior toward an individual depending on whether or not another individual is present. They may refrain from attacking an opponent in the presence of an individual who is likely to help the opponent, especially if this individual is stronger than themselves.

Further, after a fight, primates sometimes attack their enemies as a form of "revenge" or engage in friendly behavior such as grooming, which appears to be a form of "reconciliation."(12) In such cases, knowledge of the kin relationships enables the monkeys to indirectly exact revenge on the kin of the attacker rather than directly on the attacker himself/herself. Conversely, they can indirectly reconcile with the attacker's kin. Tibetan monkeys sometimes groom the offspring of the female who was their opponent in a previous agonistic interaction immediately before grooming the female herself. Further, after a fight involving their own kin, some primate species fight with, or conversely, groom their enemy's kin.(11) This can be considered as a form of indirect revenge or an indirect reconciliation. They recognize the relationships between other individuals as well as the relationship between their kin and themselves, and exhibit this knowledge during their social interactions after fights.

In an agonistic interaction, it is of great consequence if they support the wrong individual. If they are unaware of the dominant-subordinate relationships between the individuals involved in the fight as well as the relationships between other individuals and themselves, they cannot support the individual who is likely to win the fight. They cannot afford to be too lax even in cases in which both the individuals involved in the fight are weaker than themselves. Since neither individual is stronger than they are, if they support the one that is likely to lose in the aggressive interaction, the one that is likely to win the fight may garner support from his or her kin that are stronger than the individual in question, who may be defeated by that kin. Thus, if an individual does not behave appropriately in accordance with the dominance ranks in the group and the supportive relationships between individuals that are close relatives, the individual runs a high risk of being directly attacked in the agonistic interaction. Appropriate behavior in an agonistic interaction may be a result of trial-and-error learning in which the monkeys were previously attacked because of making inappropriate choices.

However, the choice of the infant males used for bridging did not necessarily exert a critical influence on the response of the recipient male. Males were rarely attacked by the recipient when they carried an infant to him. It is debatable whether they were more likely to be attacked merely because they carried an infant that was not the recipient's favorite. It is unlikely that they learned to select appropriate infants via a process of trial-and-error in which they were attacked when they carried the wrong infant. It is also possible to learn which infants to select on the basis of certain clues rather than by direct aggressive behavior. When an individual brought an infant to them, the males sometimes engaged in bridging and at other times, they did not. I examined whether they engaged in bridging less frequently when they did not receive their favorite infants. Bridging occurred in 142 of the 229 cases in which a male carried and presented an infant to another male, and in the remainder of the cases, the male ignored the infant being presented to him and did not bridge. With regard to the frequency with which the males accepted the infant, there was no statistically significant difference between the cases in which they were presented with their favorite infant and the cases in which they were presented with an infant that was not their favorite. Therefore, it may be difficult for the monkeys to learn that it is inappropriate to carry a particular infant on the basis of the recipient's responses. I believe that they associate a familiar male with his favorite infant by observing the behavior of these individuals in the course of their daily lives. In other words, the monkeys have come to understand the close relationships between infants and adults by observing that the favorite infant often accompanies, is held, and is used for bridging by its familiar male.

As seen in the case of bridging, primates alter their behavior according to the social relationships between other individuals even when they do not encounter the possibility of a direct attack or reward. Can this be attributed to the fact that they understand the concept of "relationships" in the manner that we do and because they classify the individuals in their group into categories such as "a favorite infant and his familiar male" or "a mother and her offspring?" Or, do they merely associate one individual with another separately, such as "A and a" and "B and b," and respond according to each relationship?

V. A. Dasser conducted an experiment to study the crab-eating (long-tailed) monkey, a species belonging to the genus *Macaca*. Dasser took an adult female crab-eating monkey to a laboratory, showed her pictures of the monkeys who usually lived with her in a large cage, and assessed whether she was able to identify the individuals who shared a mother-offspring relationship.[13] The subject was shown a picture of a female, a picture of her offspring (son or daughter), and a picture of another female's offspring of the same age and sex as the previous offspring. During the training stage, the subject was shown pictures of several mother-offspring pairs and was trained to select the picture of the offspring of the female. During the testing stage, the subject was shown pictures of another female and her offspring that had not been shown previously. Although the subject was presented with new pictures, she was able to identify the female's offspring, as though she had categorized her group members on the basis of "mother-offspring relationships" and had generalized this rule. She would have been confused if she had related one female and her offspring merely as "A and a" and another female and her offspring as "B and b." In this experiment, the subject, that is, the female crab-eating monkey, successfully selected the picture of the female's offspring from among two pictures. Further, in the case in which the offspring was already an adult, the subject was able to recognize whose offspring the individual was. Thus, there existed a categorization such as "a mother and her offspring" in the mind of the subject. In addition, the female crab-eating monkey was also able to recognize the differences between mother-daughter, sisters, and other related relationships.

Given the fact that crab-eating monkeys possess the ability to understand their social relationships in this manner, Tibetan monkeys, too, must understand the concept of special friendships between males and infants. In addition, they may mentally categorize various relationships such as consortships between certain males and females and supportive relationships in agonistic interactions, as well as the relationship between mothers and their offspring or between familiar males and their favorite infants.

Understanding the relationships between the self and others requires the individual to understand his or her own relationships alone. In contrast, understanding the relationships between others requires the individual to step away from his or her own perspective and recognize the social relationships between other individuals.[14] The following is one of my sister's childhood experiences at the time when our grandmother lived near us. While our mother was at work, our grandmother would come over to cook dinner for us everyday. My sister called our grandmother "grandmother" and understood that she was our grandmother. However, this was an understanding of the status of a grandmother from my sister's perspective. One day, however, when she saw our grandmother sewing our mother's clothes, she suddenly realized, "Though my grandmother is a grandmother from my perspective, she is a mother from my mother's perspective. This is the same relationship that exists between my mother and myself." In order to comprehend that the relationship between our mother and our grandmother is the same as relationship between her and her mother, my sister had to look at the relationship between our mother and other individuals from our mother's perspective and not her own. Similarly, when monkeys understand the relationships between others, they comprehend that other individuals have social relationships that are different from their own. This means that they possess the ability to step away from their own perspective and understand the social relationships between others.

Chapter 5

Studying Wild Monkeys

1. Life in the village of Fuxi

Having completed the discussion of bridging behavior among the Tibetan monkeys, I would now like to introduce my own life in Huangshan rather than that of the monkeys. Although China, during the 1980s, underwent a reformation in which the people's commune disintegrated and sightseeing by foreigners was permitted, foreigners still required a special permit in order to visit areas other than the towns and tourist resorts in China. During the spring of 1985, while I

Photo. 5–1 The author and some children in the village of Fuxi.

was still an undergraduate student, I rode a bicycle from Shanghai to Xian without acquiring permission. Since I had heard that I would be deported from China if the immigration office or security office found out about this incident, I nervously lodged at a Chinese-style hotel in the country. On one occasion, a few villagers permitted me to stay at their house. I once camped outdoors when I did not arrive at a village after nightfall. On account of these experiences, I was not excessively anxious about living in a farming village in China. However, this was my first extended sojourn in China, which afforded me a variety of experiences. I would like to introduce only a few of them in this book.

Mr. Xinming Cheng's family looked after me while I was in the village of Fuxi. Mr. Xinming Cheng's brother, Hoping Cheng, lived with his family as well as his father, Zouyuo Cheng, and his wife, in the neighboring house toward the east. In the house next to them to the west, Mr. Xinming's sister lived with her husband, Mr. Youguo. The village of Fuxi was a small farming community comprising fewer than 20 houses situated along the Fuxi River, and approximately 200 residents.

I lived in the upstairs room of Mr. Xinming's house. This arrangement was almost as though I was boarding with him. Since electricity had been introduced in Fuxi in 1989, Xinming's house had a monotone TV, but he still did not own a telephone and gas. Currently, many households are switching from the use of firewood to that of propane gas. However, during that period, firewood was burnt in the furnace for cooking purposes. People were permitted to cut and use wood that was less than 10 cm in diameter as fuel. Mr. Xinming's wife, Qiugui, prepared breakfast and dinner for me. Upon my return to Japan, I would tell my colleagues, "Of course, in China, I ate all the Chinese food that is served in Chinese restaurants daily." However, in actuality, the main dish was usually a stir-fry of radish, spinach, or greens with some tofu or pork. Each household grew their own vegetables in the farms beside their houses in the village. The corn that we gave to the monkeys was usually considered to be food for pigs. However, I requested my hosts to prepare a stir-fry of corn as a snack for me. Three fixed Chinese woks, which were about 50 cm in diameter and had no handles, were used to boil water, cook rice, and prepare stir-fries. My hosts usually ate fried rice or soup noodles for breakfast. I still crave the egg fried rice that they used to serve me for breakfast. The steamed white rice cooked in the wood furnace was also appetizing. The burnt rice at the bottom of the Chinese woks was also good. This can suffice for lunch.

Since I usually resided in the village of Fuxi from the late fall to the next spring, I received a strong impression that the climate of Huangshan is cold. In the morning during the winter, the temperature fell to about −5 degrees Centigrade. On the coldest morning, the temperature was −11 degrees

Centigrade. Around the time I would get out of bed in the morning to eat my breakfast, one of my assistants as well as Mr. Xinming Cheng would already have climbed the mountain to watch the Tibetan monkeys. I would depart from the village at 7:30 a.m., climb the mountain for an hour, and arrive at the place where the monkeys had gathered. I would conduct my observations until noon, and eat rice cooked in our observational hut for lunch. After lunch, I would, according to the Chinese custom, nap for about 30 minutes, and resume my observations in the afternoon. I would then descend the mountain and return to the village in the evening immediately before or after nightfall everyday.

Huangshan is a famous tea-producing area. The people of the village of Fuxi earn their living by cultivating tea. Huangshan Maofeng tea, and its earliest buds in particular, are considered to be one of the best tea leaves. The tea leaves are picked during the entire month of May. During this period, elementary schools are closed and the children, too, join in the tea picking. When the people's commune disintegrated in the 1980s, the tea plantations were redistributed to each household. The villagers would begin picking the tea early in the morning, and would process the tea leaves until late at night. They grilled the tea leaves in the wok that they usually used to cook rice, and dried the leaves by rubbing them with their hands. The whole village becomes extremely busy during this season. Mr. Xinming made a joint investment with neighbors in a machine to grill the tea leaves. They began using this machine once the tea leaves grew to a certain extent. The villagers also hire part time workers from adjacent villages. Fortunately, since monkeys do not eat tea leaves, the residents of the village of Fuxi and the monkeys never get into confrontations over the tea leaves.

However, they do compete for bamboo shoots. Since the bamboo shoot season overlaps with the busy season during which the tea is picked, the people of the village do not have an advantage over the monkeys. During this season, the Tibetan monkeys sometimes forage for bamboo shoots deep within the bamboo thickets. Since the people are busy picking tea and the monkeys often go far over the mountains to search for food in the woods, the observations of the monkeys must be interrupted during this period.

My observations were interrupted during the period beginning from April 24, 1991, when I went through Urumqi to Kaxgar. However, when I returned to the village of Fuxi and attempted to resume my observations on May 16th, we were unable to locate the Yulingkeng group. Mr. Xinming and I searched for the monkeys deep in the mountain. However, many leeches emerge from the mountain during this season, and innumerable leeches entered my socks or pants during one day walk everyday. While the leeches were tolerable, we also had to be careful of snakes during the warm season. Green, slender snakes known as "Zhuyequng" (*Trimeresurus stejnegeri*) inhabit Huangshan. These snakes produce nerve venom. We once found a snake twined on the branch near the observational hut, and although Mr. Xinming smashed its head and killed it, its body continued writhing. Mr. Ming Li persistently continued hitting its head. An even more formidable snake than Zhuyequng is *Agkistrodon acutus*, which is referred to as "Wubulong", since the people who are bitten by it die before taking five steps ("Wu", "bu", and "long" mean "five", "step", and "dragon" in Chinese, respectively). This snake produces hemolytic venom. I once saw this snake being captured by a villager. The snake had a very large head. The reason the villager made the effort to capture such a dangerous snake is that these snakes can be sold at a very high price.

I finally stayed in the village of Fuxi until June 5, 1990, but ultimately abandoned the search for the Yulingkeng group of that year and returned to Japan.

Thus, my observations had to be discontinued during the season for tea picking. In the winter, however, Mr. Xinming informed me that he could continue to assist me in my observations of the monkeys during the Chinese New Year's Days. I stayed in Fuxi on the Chinese New Year's Days in 1992. On this day, people slaughter the pigs that they have raised for a year; thus, large quantities of pork intestines are served. The slaughtered pig is then hung from the ceiling and the required amount of meat is cut at each meal.

The pork served at the dining table consisted of mostly the fat of the pig. While I was fond of fatty meat, it was rather difficult to eat meat that consisted of fat alone. I could eat fatty meat if it contained at least a small proportion of meat. However, since I was boarding with my hosts, I sometimes ate the meat even though it was purely fat. Later, I realized that Xinming's children did not eat the parts that consisted solely of fat but carefully chose the parts that had more meat. Mr. Xinming and I were the only ones who ate the parts that were purely fat.

The people of the village usually drank distilled spirits that are known as "Baijiu" in Chinese. At parties in China, the custom is to drink the alcohol poured into the cup and say "Cheers!" At this point, the hosts and guests have to invert the cup after they have finished drinking to prove that they have emptied it. Moreover, this practice may sometimes be enforced on all the guests at the party. While I was in the village of Fuxi, I usually drank at my own pace. On the New Year's Day, however, I fell sick on account of drinking too much because I had followed this Chinese custom at the party in Xinming's house.

When we completed our observations each year, we usually returned to Anhui University for a farewell party. I was concerned about the fact that we had one party for lunch and another for dinner in the same day. The guests at both parties were largely the same, that is, the research colleagues studying the

Huangshan monkeys and the staff of Anhui University. However, there were two parties because there were two different hosts. After drinking and eating heartily at the lunch party, after an interval, I returned to the table for drinks at the second party. At this point I thought, "I was beaten by China."

2. Monkeys at Emeishan

Emeishan (Mt. Emei) is the other mountain besides Huangshan where Tibetan monkeys can be easily observed. Emeishan is located 150 km southwest of Chengtou, the capital city of Sichuan province. Emeishan is one of the four spiritual centers of Chinese Buddhism. The altitude of this peak is 3,099 m. Like Huangshan, this mountain can be scaled on walking trails made of stone. Tibetan monkeys can be encountered along these trails. The monkeys at Emeishan do not flee from humans; rather, they approach tourists in order to beg for food. I assume that in the past, the monks at a temple on Emeishan gave them leftover food. If such provisioning has been undertaken for a few hundred years, this may be one of the oldest provisioned group of monkeys. The recent increase in the number of tourists has led to the emergence of the current situation.

Ming Li and I arrived at Emeishan on May 28, 1990, in order to observe the Tibetan monkeys. We stayed in Kunming and met Mr. Qikun Zhao from the Kunming Institute of Zoology. He has been studying the Tibetan monkeys at Emeishan and has published extensive articles about them.[(1) (2) (3) (4) (5)]

At Emeishan, we passed through Jieyindian via Qingyingao, Hongchunping, Xianfengsi, Yuxiansi, and Xixiangchibo, and headed toward Wanfoding, which is situated at the top of the mountain. We then came down to Qingyingao through Wanniansi.

During these five days, we saw six groups of Tibetan monkeys. Unfortunately, we were unable to witness an encounter between any two groups. Yet, Emeishan provided us with the opportunity to observe multiple groups. Unfortunately, however, the Tibetan monkeys at Emeishan were congregating beside the path to wait for the sweets or bread that was given to them by the tourists. This situation resulted in several conflicts between the tourists and monkeys.

3. Problems in provisioning the monkeys

Determining the manner in which monkeys and humans should interact is a difficult task. This problem is not confined to Emeishan, and the Yulingkeng group at Huangshan is not exempt from this problem. As previously mentioned, the Yulingkeng group was first provisioned in 1985. Since then, the individual monkeys belonging to this group have been identified and research based on individual identification has made progress. When Dr. Kazuo Wada first began the study of Tibetan monkeys at Huangshan in a joint research project with Anhui University, a telemeter was fixed to Wushang's neck so that the researchers could locate the Yulingkeng group easily and thereby observe Tibetan monkeys in the wild. However the steep slopes and cliffs of the mountain at Fuxi made it very difficult to follow the monkeys. Therefore, the researchers attempted to carry out provisioning by scattering corn. While the amount of corn that was scattered was changed every year depending on the number of monkeys in the group, it usually amounted to a total of 6 kg per day. After 1985, we provisioned the Yulingkeng group only during the several months in a year during which we observed the monkeys, and they returned to the wild when our study period was complete. When we began the next study, we looked for the monkeys again. This situation continued for several years.

However, the 1990s saw the widening of the gap between the life of the people working at Huangshan, which is visited by many tourists, and the life of the people in the village of Fuxi, who earn their living by picking tea, although the distance between the two areas is not much. One of the various differences between these regions is their economical condition. In 1994, the Huangshan Management Bureau, the forestry bureau in Tangkou, and the people in the villages of Zhaixi and Fuxi formulated a plan to promote the Tibetan monkeys as the sightseeing attraction of Fuxi in order to attract the tourists who visit Huangshan. I have been told that the Huangshan Management Bureau attempted to provision a group of Tibetan monkeys in 1986; however, their plan failed because the monkeys were not enticed by the food. The Management Bureau then attempted to establish a monkey park, and to this end, they decided to use the Yulingkeng group that had already been provisioned in Fuxi. I observed the Tibetan monkeys belonging to the Yulingkeng group until 1993. My final visit to the village of Fuxi overlapped with the period during which the road from the village of Zhaixi to the village of Fuxi was being reconstructed for the establishment of the monkey park. Since I have not visited China since the construction of the monkey park, I have not had an opportunity to observe the progress of the Tibetan monkeys in the Yulingkeng group. I have been informed that, so far, the monkey park in the village of Fuxi has been thriving. If tourists are allowed to feed the monkeys, the monkeys will become too accustomed to humans and man-made food and attack the tourists to obtain the food. Therefore, in the park in Fuxi, the tourists are not allowed to feed the monkeys and must maintain a certain distance from them. This ensures that the tourists enjoy watching the monkeys.

However, in the future, this park may encounter problems related to the population growth of the group due to provisioning. If the quantity of food provided to the monkeys is decreased when their population increases, the monkeys, who are accustomed to humans, may harm the crops. In addition, the male monkeys who remember the taste of food and are accustomed to humans could move to another group and encourage the monkeys of that group to damage the crops. In fact, rhesus monkeys, which live closer to the fields than Tibetan monkeys, damaged the crops in a farming village near Fuxi. Since rhesus monkeys, too, are a protected species of the second level, people are prohibited from capturing them; as a result, the monkeys are a nuisance to the people whose fields they damage. Since the quantity of food provided to the monkeys cannot be decreased, the population of the provisioned monkeys will increase if the number of individuals in the group is not regulated. Reducing the number of monkeys poses another problem. After encouraging an increase in the number of monkeys for the benefit of the people, is it acceptable to take the lives of newborns for the convenience of the people? Regardless of how difficult its management becomes, the monkey park cannot be abandoned. I hope that by learning from the challenges faced by Japanese monkey parks in the past, the people of Fuxi will be successful in creating an environment in which both monkeys and humans will be able to coexist in the forest of Huangshan.

Chapter 6

The Lives of the Males and Females

1. From mating to birth

I will now return to the topic of primates and their social relationships and discuss the lives of male and female Tibetan monkeys. The vigorous copulation among Tibetan monkeys—as is evident in the case of Huangmao—is impressive. During the fall, I heard the voices of the monkeys that were copulating in the woods everyday. The Tibetan monkeys continued copulating in January, and the females sometimes engaged in mating activity even while they were pregnant. The females that did not give birth to a baby in the spring sometimes engaged in copulation during this period as well. Baijamao, who did not give birth in the spring of 1990, continued copulating in April, and the male would ejaculate after copulating with her. However, the females rejected copulation with any male during the months in which they gave birth to babies.

Incidentally, there are few nonhuman primates among whom the males force females into copulation. If the females reject these attempts at copulation by placing their buttocks on the ground or running away, even the strong males have to abandon their attempts at copulation.

Therefore, the only option available to the strong males is to intervene in the mating activity between subordinate males and females. The subordinate males are aware of this, and they do not attempt to engage in copulation while they are within the sight of the dominant males. Although the subordinate males have opportunities to copulate with the females in the woods, they could, unfortunately, get caught by the dominant males because of the vocalizations of the females during copulation. Since Tibetan monkeys maintain a smaller distance between individuals than Japanese monkeys, it is difficult for the males to find opportunities to be alone with the females. On one occasion, a low-ranking male, Yanbaiban, arrived at the provisioning site before the other males

Photo. 6–1 Genital inspection.

and attempted to copulate with a female. However, he immediately looked back and terminated the copulation; he may have realized that the dominant males would soon approach from their usual direction. When the dominant males found a subordinate male copulating, they often attacked the female. At times, when the alpha male found the second-ranking male copulating with a female, the alpha male did not attack the male but engaged in copulation with the female.

Female chimpanzees have pink genital skin, which is swollen for several days around the time of ovulation. This is extremely noticeable in the woods. The genital skin of female Japanese monkeys becomes red during estrus. The belief that "a Japanese monkey's butt is red" is derived from this fact. In contrast, Tibetan monkeys do not display any apparent signs during ovulation. Through their frequent inspection of the genitals of the females, the male Tibetan monkeys may be attempting to monitor the estrus cycles of the females. The males observed, sniffed, licked, and touched the genitals of the females with their fingers, and then licked their fingers (Photo. 6–1).

Although male and female Tibetan monkeys perform various actions before copulation, the act of copulation itself is very simple. When the male mounts the female, he inserts his penis and thrusts a few times until he ejaculates.(1) However, it is said that the Tibetan monkeys at Emeishan, especially the high-ranking males, occasionally ejaculate after repeated mountings. In this case, as in the case of Japanese monkeys, the male inserts his penis, then dismounts several times before ejaculation. Male Tibetan monkeys sometimes continue mounting a female and inserting their penises even after ejaculation. This creates quite a commotion because several monkeys gather to disturb the copulation of the male who, despite being a high-ranking male, cannot attack the others until he finishes copulating. After copulation, the male usually sits down and eats the sperms adhering to his penis, and the female often grooms him. The copulation that the alpha male, Huangmao, engaged in was so simple that I was inclined to believe that he mated in order to have the female groom him.

Tibetan monkeys are promiscuous, and during the mating season, the males copulate with multiple females, and the females, with multiple males. However, this does not mean that they mate with each other randomly; they have a preference for certain mates. In addition, I observed that during the mating season, a male would follow a particular female for a certain period of time, thereby giving the impression that he was in love with her. I will refer to this male-female relationship as a "consortship." However, since the females were usually around the other members of the group, the male and the female were rarely allowed complete privacy. In addition, the females did not direct any remarkable sexually active behaviors toward their consort males any more than they did toward the other males. Therefore, this relationship is slightly different from the general "consortship" among other primates. Fig. 6–1 presents the proximity between a male and a female during the consortship period. The male spent 82% of his time within 2.5 m from his consort female. This data, which does not include the observations conducted at the provisioning site or the sleeping site, indicates that the male remained in 2.5 m proximity to the female in the woods. I used the Hinde Index to examine whether it was the male or the female who maintained this proximity. The Hinde Index is calculated by [{(the number instances of male approach—the number of instances of female approach)/(the total number of approaches)}—{(the number of instances in which the male leaves—the number of instances in which the female leaves)/(the total number instances in which an individual leaves)}]. For example, if a male approaches a female seven times and a female approaches a male three times, the proportion of the male approach is 0.7, and if a male leaves a female four times and a female leaves a male six times, the proportion of the instances in which the male leaves is 0.4; then, the Hinde Index would be 0.3 (0.7−0.4). The Hinde Index ranges from −1 to +1. A value that is positive and close to 1 implies that the male approaches the female and does not leave her. Thus, it is the male who are responsible for the proximity between the two individuals. On calculating the Hinde index for the Tibetan monkeys in a consortship, I obtained, as I expected, a positive value. In addition, the value was higher in the case of the males and females who were in a consortship than in the case of the males

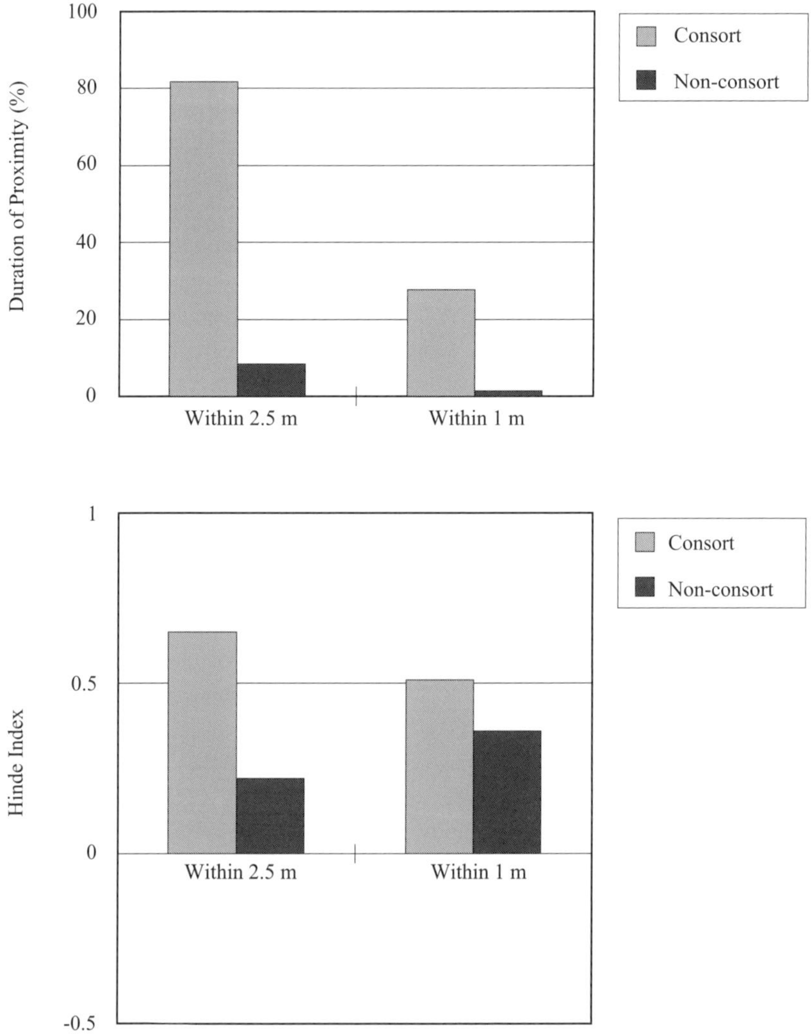

Fig. 6–1 Proximity of a male-female pair in a consortship.
Within 2.5 m: A male and a female were within 2.5 m of each other. Close Proximity: A male and a female were within 1m of each other. Hinde Index: [{(the number of instances of male approach—the number of instances of female approach)/(the total number of approaches)}—{(the number of instances in which a male leaves—the number of instances in which a female leaves)/(the total of number instances in which an individual leaves)}].
Consorts: A male and a female were in a consortship. Non-consorts: a male and a female were not in a consortship.

Photo. 6–2 A male and female in a consortship.

and females who were not in a consortship, although the average indcx was positive in all male female pairs. However, with regard to close proximity, that is, 1 m distance between the two individuals, the value was not very high, and the index calculated during consortship was not significantly higher than the index calculated while the monkeys were not in a consortship. Therefore, it may not be necessary for the male to remain very close to the female, that is, close enough to make physical contact; however, he tries to remain within sight of her.

While it was generally the males who followed their consort females, the females, too, sometimes approached the males and frequently groomed them. The male and female formed a very intimate relationship during the cosortship (Photo. 6–2). This relationship usually lasted for several days, weeks, or even months, but was not evident during the non-mating seasons.

The higher-ranking males formed consortships with females more often than the lower-ranking males. During the period from 1989 to 1992, I observed 45 consortships among the Tibetan monkeys. While 12 of the 14 higher-raking males formed a consortship at least once, only two of the 14 lower-ranking males formed a consortship.

Why did the high-ranking males engage in consortships? On one occasion, for example, the alpha male and his consort female did not appear at the provisioning site for an entire day. They may have been developing a relationship with each other away from the group. During this time, however, the other males, especially the second-ranking male, copulated with several females.

It seems probable that low-ranking males also benefit from consortships with females. In order to examine this possibility, it is necessary to investigate the estrus cycles of the females and the extent to which the males monitor the female sexual condition by means of genital inspection.

Unlike in the case of humans, young, female nonhuman primates who have never had offspring are generally not popular among the adult males. During my observations in 1989, however, I received the impression that young, female Tibetan monkeys were fairly popular among the adult males. Although Beihen and Tougou were only three and a half years old, Huangmao and Wushang sometimes waited to mate with them while they played with the other infants and juvenile monkeys. This behavior of the adult males appeared funny to me. However, this may be the result of a series of deaths that occurred in the Yulingkeng group in 1988, because of which the number of adult females decreased. After 1990, the juvenile females were often observed to associate with young males who were not yet completely mature. For example, when Sanjiaolian was four years old, she would accompany Chawei, who was older than her by one year, and often mated with him. Further, females of this age often copulate with males who are younger than themselves. I was astonished to observe Huagang, a male who was less than three years old, mating with a female and actually ejaculated.

A characteristic of Tibetan monkeys is that they engage in copulation, culminating in ejaculation, in various social contexts. Male and female Tibetan monkeys mate even in the social context in which, if they were males, they may have performed friendly social behaviors such as penis sucking and mounting. For example, in the case of aggressive interactions in which one female attacked another, a male would approach the aggressor and copulate with her. The males might attempt to calm her down. It was amazing that in such social contexts, male and female Tibetan monkeys engaged in copulation until the male ejaculated.

2. Males, females, and infants

I will now turn to the social relationships between males, females and their infants.

In certain primate species, the males have few direct interactions with infants in the group, whereas in other species, the males engage in frequent intensive interactions with infants and help the mother raise the infants. Since the males cannot nurse the infants, paternal care usually comprises grooming and carrying them. The males also protect the infants from predators, monkeys belonging to other groups, and other members of their own group. Though the males contribute toward survival and growth of infants indirectly, only their direct interactions will be discussed. The quality and the number of the interactions between males and infants of several primate species have been studied and compared.[(2)] One of the factors influencing the degree of paternal care is the certainty of paternity. It would be a waste of time and energy for a male to devote himself to the care of an infant that could be the offspring of another male. In a monogamous group (one-male, one-female group), it is reasonable for the male to look after the infant since the paternity of the infant is relatively clear. However, males in a one-male, multi-female group rarely demonstrate paternal care. Yet, despite the possibility that the females will copulate with males other than the resident male of the group, the paternity in a one-male, multi-female group is relatively certain. In contrast, the males of certain species interact with the infants in a multi-male, multi-female group although these groups make it more difficult to determine the paternity of the offspring.[(3)] [(4)] One species that displays this type of behavior is the Barbary monkey, which I have discussed in this book on account of its bridging behavior. As in the case of Tibetan monkeys, it has been reported that male Barbary monkeys take special care of particular infants. The nature of this intimate relationship between males and infants among Barbary monkeys is disputed. As already mentioned, Taub assumed that the monkeys that shared this relationship were matrilineally related. However, when A. Paul and J. Kuester conducted an investigation, they discovered that male Barbary monkeys did not necessarily have more intimate relationships with the infants that were matrilineally related to them. The observations on their copulative behavior in the previous mating season revealed that males did not necessarily share a close relationship with their possible offspring. In addition, a DNA analysis of the infants and males revealed that the adult males did not necessarily like to look after their own offspring.[(5)] Some researchers hold that Barbary monkeys take very good care of the infants because the females mate with so many males in their group that the paternity of the infants is uncertain. The females attempt to convince each male that the newborn is his offspring; therefore, since the mothers know that the males will not harm their infants, they tolerate the male care directed toward their babies.

Are male Barbary monkeys tricked into looking after the infants? It is not critical for an infant less than one year old to have a friendly relationship with a particular adult male in order to ensure infant survival. While the males of species such as the Marmosets of South America look after their own descendants or relatives because this is actually advantageous to the survival of the infants, infant handling by males in a multi-male, multi-female group may have other implications.

Why do the adult males in a multi-male, multi-female group engage in friendly interactions with infants? One possible reason is that this is an attempt to attract the females. B. Smuts studied savanna baboons and reported that adult males develop an intimate relationship with the infant of the female whom they wish to mate with.(6) It is also held that male gelada baboons attempt to solicit support from the mother of an infant by holding the infant in agonistic interactions with other males.(7)

How are the similar interactions observed between males and infants among Tibetan monkeys? Which infant did the males favor, hold, and groom? I analyzed 173 pairs of males and infants; the latter were born between 1989 and 1992. I will not discuss the details of the study; the results, however, are as follows.

First, the males did not take special care of their maternal siblings.

Second, I did not find any evidence that they favored their own offspring. Unfortunately, since I did not examine the DNA of the Tibetan monkeys in the Yulingkeng group in order to confirm their paternity, I could not identify the biological father of each infant. However, the juvenile or new immigrant males that had not mated with adult females during the previous mating season often held infants during the subsequent season; these males had their favorite infants. This indicates that males often hold infants who are not their offspring. Although the Tibetan monkeys formed consortships, they did not follow one particular female, but followed several different females throughout the mating season. As in the case of Barbary monkeys, since the females mated with multiple males, any male could be the father of the offspring. The males interact with the infants because the mothers are tolerant of infant handling by the males. This may be because the mothers know that since the males are aware that the infants could be their offspring, they will not harm them.

The final possibility is that the males may favor the infant of a female they are attracted to. The intimate relationships between the males and the infants among Tibetan monkeys are usually determined by the male's future social relationships with the infant's mother or his already established consortship with her, rather than paternity or kinship.(9) In some cases, the males favored the infants of their familiar females.

In 1991, for example, Duyanlong entered into an extended consortship with Beihen. Since Beihen did not have any offspring at this time, the situation was simple, that is, Duyanlong liked Beihen. In 1992, however, Duyanlong formed a consortship with Tougou. He often held and groomed Toutai, Tougou's daughter, and used this infant for bridging with other males. In other words, Duyanlong favored the offspring infant of his consort female, and held and used the infant in bridging. On the other hand, Duyanlong never held Beixiao,

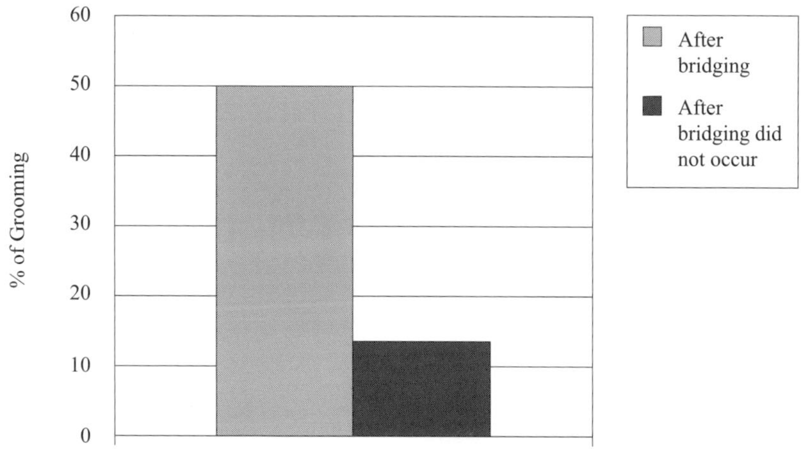

Fig. 6–2 Grooming between a male and a female after bridging.
The percentage of social grooming between a male and a female after bridging did and did not occur. After bridging: Bridging occurred when either a male or a female presented an infant to the other. After bridging did not occur: Bridging did not occur when one of the two rejected the infant presented to him or her by the other.

Beihen's infant, regardless of how often the infant approached him. He was not concerned about the fact that this was the baby of Beihen, whom he had been very close to the previous year.

Did the males hold the offspring of the females with whom they formed a consortship? A statistical analysis, unfortunately, did not reach significance. However, I obtained the following results by examining the data on bridging between the males and females.

During my study periods between 1989 and 1992, excluding the birth seasons in the spring; Dec. 5, 1989 – Jan. 21, 1990; Dec. 12, 1990 – Jan. 11, 1991; Dec. 5, 1991 – Jan. 15, 1992; Sept. 5, 1991 – Nov. 4, 1991, I recorded a total of 54 cases of bridging behavior following 113 interactions in which a male and a female almost bridged, that is, either one of them held an infant and presented it to the other. Only four (7.4%) of these 54 cases occurred in a tense situation, such as after an agonistic interaction in the group. The rest of the cases occurred during situations of calm that were not preceded by any noteworthy encounter. There was more grooming between the male and the female after a successful bridging than after an unsuccessful attempt to bridge (Fig. 6–2). These observations also hold true for bridging between males.

However, the following observations present the contrast between the bridging behavior between males and females and that between males. In cases of bridging between the former, the male initiated the behavior more often than

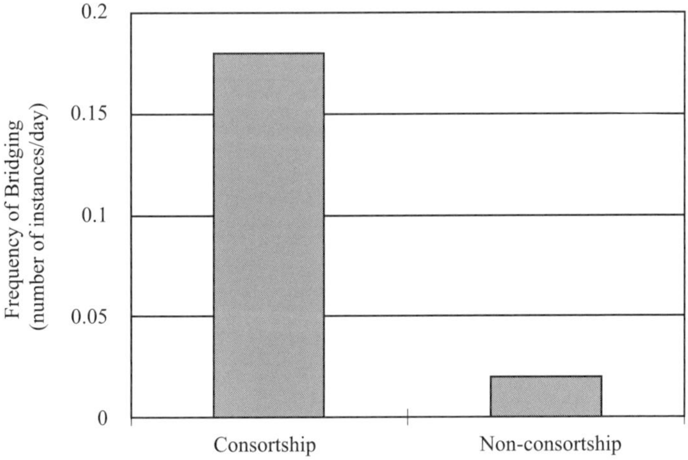

Fig. 6–3 Bridging during consortship and non-consortship.

the female, that is, in 71 (62.8%) cases out of the total observational counts. The male brought an infant to the female most frequently, that is, in 52 (46.0%) of the total observations, and actual bridging occurred in 15 of 52 interactions. Higher-ranking males frequently carried an infant to a female and engaged in bridging behavior with females more frequently than the lower-ranking males. Moreover, these higher-ranking males brought an infant to the female who was in a consortship with them. During the consortship, the males brought an infant to their consort female and engaged in bridging behavior more frequently than they did with the same female during the period in which they were not in a consortship (Fig. 6 – 3). These higher-ranking males were undoubtedly dominant to all the females; therefore, they were rarely attacked by the females for merely not presenting them with an infant. The high-ranking males may have presented an infant to their consort females in order to form an association with them.

With regard to the infants that were used in this kind of bridging, male infants were not used more frequently in bridging between males and females than female infants, while male infants were used for bridging between males. The males usually presented the female with her son or daughter (Fig. 6–4). The males might use the offspring of the female in order to better relate to the female. Here, the hypothesis "My friend's friend is my friend" can be applied again. Another Japanese proverb may be more suitable to this case: "You should shoot an arrow to the horse before aim at the soldier on it."

In the cases of bridging between the males, whether or not the infant was the males' favorite did not result in any difference in their response, that is, engaging or not engaging in bridging. In contrast, an examination of the bridging

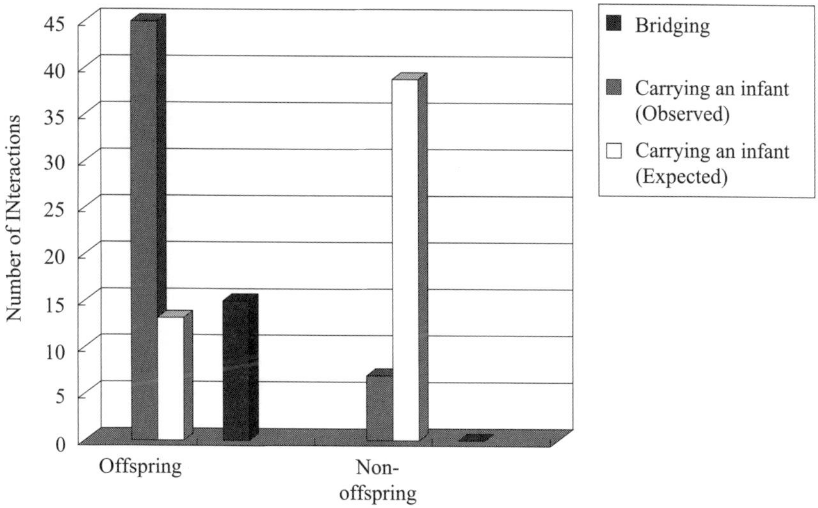

Fig. 6–4 Offspring and non-offspring infants used in bridging behavior.
The number of interactions in which infants who were and were not offspring were used.

between males and females reveals a difference in the response of the female, depending on whether or not the infant was hers. The females engaged in bridging behavior with males more frequently when they were presented with their own offspring than when they were presented with another individual's offspring (Fig. 6–4).

If the females did not have any offspring, the males were compelled to use another individual's. For example, Ye did not give birth to an infant during one particular year. However, Duyanlong dealt with this situation by bringing Ye her granddaughter. This infant was the daughter of Tougou, who was Ye's daughter. There were two cases in which a female who had granddaughters was presented with infants other than her offspring; both of these were instances in which Duyanlong brought Ye her granddaughter. This appeared to be an intelligent choice despite the fact that he was unable to bridge with Ye.

From the above results, it is possible to conclude that bridging between a male and a female is a behavior in which the relationship between the two individuals is manipulated by using an infant as a third party. The males establish or maintain a friendly relationship with a female by using her offspring.

D. Premack trained a chimpanzee named "Sarah" in language acquisition. In a series of experiments, Sarah indicated that the relationship between "a bottle opener and a bottle" is the same as that between "a key and a door," but that these two relationships differ from the relationship between "a pair of scissors and paper."[10] This suggests that the chimpanzee possessed the ability to understand the correspondence between one relationship and another.

Tibetan monkeys exhibited a clear difference between the way the males used their partner's favorite infant during bridging with other males and the way they used the female's offspring while bridging with the female. While the males sometimes presented their male partner with an infant who was not his favorite, they rarely carried an infant to a female who was not its mother. I present this difference with caution because there are many other differences in the bridging between males and that between males and females. However, I wonder whether Tibetan monkeys can differentiate between the relationship of an infant and its mother and that of an infant and its familiar male. This differentiation may depend on the difference in the frequency of the social interactions between a mother and her offspring infant and those between a male and his favorite infant. However, I would like to recall the experiment conducted by Dasser, in which a crab-eating monkey pointed out the mother-offspring relationship in the group. Primates can recognize a mother-offspring relationship even after the offspring infant becomes an adult and does not engage in frequent social interactions with its mother. The behavior of male Tibetan monkeys differed not only according to the difference in the frequency of interactions between a mother and her offspring infant and those between a male and an infant, but also according to the monkeys' differentiation between the mother-offspring relationship and the close relationship between a male and his favorite infant. Therefore, the males often used the favorite infant of their male partners and always used the offspring of their female partners.

Infants that were not yet one year old were the ones that were most frequently held and used in bridging between males; this is because of the infant's characteristics such as age, sex, body size, and so on. In addition, the frequency with which the infants were used in bridging was influenced by the social relationships between the infant and the male that received it. This is one of the social characteristics of the infant. Similarly, the rank of the mother influences other individuals' response to her offspring.

There exists a further possibility. Assume that the alpha male is in a consortship with a female, and he favors her offspring. In order to bridge with the alpha male, the other high-ranking males will present him with the infant he favors. Further, the other males that observe this behavior may attempt to bridge with these high-ranking males using the same infant. This may continue even after the alpha male terminates his consortship with the female. In this case, although it is no longer necessary for the males to use this particular infant, the infant will still be used frequently in bridging between the males since, having been used in bridging in the group so far, the infant now carries a certain social value.

This is similar to the use of currency in human life. Since bills can no longer be exchanged for gold, their value is supported by the members of society who believe that these bills have a certain value. Would it be an exaggeration to say, "It is not humans alone who understand the concept of social value, but Tibetan monkeys as well"?

The reader may wonder whether monkeys carry a favorite infant to its familiar male because they think, "My partner thinks that this particular infant is precious," and whether they understand another individual's mind so as to say, "This individual thinks this." I would like to discuss this topic in the last chapter (Chapter 10).

3. From infancy to adulthood

Infant Tibetan monkeys are often born during the period from April to May. Tibetan monkeys, like Japanese monkeys, inhabit an area that experiences four distinct seasons. During the winter, there is a scarcity of food in the woods, but during the spring, the monkeys eat new buds, young leaves, and flowers; they may also feed on bamboo shoots. During the fall, there are plenty of fruits and seeds such as acorns. The monkeys give birth to babies during the spring in order to ensure that they are old enough to survive by the winter, when there is little to eat besides the barks of trees. The infants that are more than eight months old may be able to survive the harshest season of the year.

The gestation period of Tibetan monkeys is approximately six months, which is the same as that of other closely related macaques. In order to give birth to their babies in the spring, the monkeys must mate during the fall. They have a prolonged mating season that extends from late summer to winter. Generally, however, the fall is their mating season and spring is their birth season. The Tibetan monkeys in the Yulingkeng group had a high birth rate, though not necessarily on account of their frequent mating.[(11)] On an average, the adult females gave birth once in 1.36 years. The provisioning of the monkeys may have increased their birth rate. However, although the females were provided with corn for only about four months during the study, their birth rate was almost the same as that of Japanese monkeys in captivity or in a fully provisioned group.

Like stump-tailed monkeys, newborn Tibetan monkeys are light in color. While newborn stump-tailed monkeys are white, newborn Tibetan monkeys are of a light cream color. This color dulls within six months. As is indicated by Huangmao's name ("Huang" means "yellow" and "mao" means "fur" in

Chinese), young adult males have bright-colored fur, which darkens as they grow older. The females do not undergo a change in color that is as conspicuous as that of the males. Instead, their faces, especially their upper halves, become red as they grow to adulthood.

Immediately after the females had given birth, bridging, especially between those who had recently had babies, took place. This behavior amused me as it appeared to be a celebration of the birth of the babies. After the babies reached the age of six months, young, nullparous females would sometimes approach the mother holding an infant, and engage in bridging. However, the mothers guarded their babies for a certain period after their birth. The other females, especially the low-ranking ones, were unable to touch the babies unless they flattered the mothers by grooming them. In contrast, after the baby reached the age of six months, a female that had a fight with the mother would groom the infant first, and then groom the mother. That is, the order of the situation was reversed.

The babies gradually began to leave their mothers and play with other infants. The infants between the ages of six months and two years were very attractive to me, and I was often tempted to approach them. When I began my observations, however, Wushang would usually interpose between me and the infants and stare at me. The infants would become upset if I came too close to them, and Wushang would then approach me in a threatening manner. He was protecting the infants. However, he never attacked me. He pounded the ground with his hands or shook his head with his mouth open, and sometimes cried out, making a noise that sounded like "Gwau." The infants, too, displayed the same behavior and threatened me. They would grab one of the branches of the nearest tree, pound the ground, or threaten me with their mouths open, making a sound that resembled "Gao." This scene, too, was amusing. However, Wushang did not provide any special protection to his favorite infant against attacks by the other members of the group.

Is being a male's favorite beneficial to an infant? It was clear that a larger number of male infants had close relationships with adult males than did female infants. If the male infants did obtain certain benefits from their relationship with adult males, then the mortality rate of the male infants should have been lower than that of the female infants. However, there was no significant difference between the two rates. Yet, this was not an adequate comparison because there were few instances of infant deaths. It may be easier for the mothers to raise sons than to raise daughters if several adult males look after male infants. If this is true, the female would have a shorter interbirth period after the birth of a son than she would after the birth of a daughter. After the birth of a son, on an average, a female had her next baby after 1.27 years, and if she had a daughter, she had her next baby after 1.43 years. This data does not reveal any statistically significant difference. Ultimately, I gathered no data supporting the hypothesis that having a relationship with adult males is beneficial to the infants. Tibetan monkeys engage in infant handling not in order to take care of the infants but to use them as social tools for bridging in order to establish or maintain their relationships with other males, or to associate with females, as in the case of Barbary monkeys.

However, infant Tibetan monkeys are not merely used in bridging, but also indirectly benefit from the bridging behavior.

Barbary monkeys and Tibetan monkeys differ with respect to the period during which the infants are used in bridging. Barbary monkeys frequently use their infants in bridging for some time after their birth; however, the frequency of this decreases as the infants grow older. On the other hand, infant Tibetan monkeys are often held and used in bridging even after they are six months old. This period coincides with the peak of the mating season, and therefore, the males use the infants to approach the females as well.

Male infants were used in bridging more frequently than female infants. As a result, male infants had frequent opportunities to interact with adult and juvenile males. The male infants also had more opportunities to groom the adult males between whom they were caught after bridging. Males who were two or three years old often played with other individuals of the same age and often engaged in bridging while playing with one another. However, this is not the only behavior they engaged in. Male infants often approached adult males in order to bridge with them. The males maintained their relationships with adult males by being used in bridging when they were infants, and by sometimes bridging with adult males before this relationship ended, that is, while they were still being used in bridging (Table 6–1). In addition, juvenile Tibetan monkeys frequently carried an infant to and had interactions with high-ranking males. Juvenile Japanese monkeys, on the other hand, would have been compelled to remain at the periphery of the group. However, juvenile male Tibetan monkeys maintained a social relationship with the adult males. The juvenile males may have been able to maintain their relationship with the adult males because they had continuously interacted with these adults by being used in bridging from infancy to the time they were infants and by bridging from the time they were infants to the time they were juveniles.

In addition, the juvenile males Duyanlong and Sanxiong, who had frequent interactions with the adult males in the group, later became the alpha males. Duyanlong was the alpha male during the winter of 1990, and Sanxiong during the winter of 1992. I suspect that Huangmao, the former alpha male, may have been born in this group. These alpha males did not emigrate from the Yulingkeng group, possibly because this group was provisioned. However, all

Table 6–1 Development of social behavior in males and females.

Age (years)	Male		Female	
	Interactions primarily with adult males.	Interactions primarily with adult females.	Interactions primarily with adult males.	Interactions primarily with adult females.
0	Begin being held and used in bridging by males.	Begin being held and used in bridging by females.	Begin being held and used in bridging by males.	Begin being held and used in bridging by females.
0.5		Begin mounting females.		
1			No longer held and used in bridging by males.	No longer held and used in bridging by females.
1.5			Begin performing presenting behavior to males.	
2	Begin holding infants and bridging with other males.			
2.5				
3	No longer held and used in bridging by males.	No longer held and used in bridging by females.	Begin holding infants.	
3.5	Emigrate from the natal group (in some cases).	Begin copulating with females, culminating in ejaculation.	Begin copulating with males. Enter into consortships with males.	
4			Begin bridging with males.	
4.5		Enter into consortships with juvenile females.		
5			Begin giving birth.	Begin bridging with other females.
5.5				
6	Become the alpha male of the natal group (in some cases).			
6.5		Enter into consortships with adult females.		
7				
8+				

Photo. 6–3 Presenting behavior.

juvenile males remained in their natal group until they became adult. In addition, the immigrant males in this group share certain common characteristics. During my study, most of the males who immigrated into this group were adults, as seen in the case of Chiyan and Yanbaiban, who were adults in 1989. The only exception to this tendency was observed during the winter of 1991, when several juvenile males immigrated into this group. However, their case was unique. Further, at Huangshan, I never observed a solitary male outside the group. The fact that this group had an almost equal number of adult males and females may indicate that male Tibetan monkeys remain solitary for a very short period of time. Tibetan monkeys may remain in their natal groups until maturity and, in some cases, until they become the alpha males. They may then emigrate from their natal groups, spend a very short time as solitary males, and immigrate into another group. The life history of male Tibetan monkeys, including the period during which they shift between groups and the way they do so, may be related to their social behavior and the characteristics of their social relationships. This makes me wonder how the male Tibetan monkeys at Emeishan, where there are several provisioned groups, shift from one group to another.

On the other hand, how do the interactions with other individuals affect the growth of the females? After female infants reached the age of one year, they were no longer used in bridging; rather, they often began holding infants themselves. Unlike the males, however, the females did not engage in bridging behavior either with other adults or with individuals of their own age until they reached adulthood.

At this age, the females, for the most part, interacted with their own mothers and sisters.

When they reached the age of two years, the infants, like the adult females, attempted to interact with the adult males. Although Keai, who had just reached the age of two years, was not treated as an adult, she began to approach the adult males and perform the presenting behavior in a manner similar to the adult females (Photo. 6–3). Although she was not yet able to copulate, Huangmao inspected her genitalia. She then appeared satisfied and groomed him.

Tibetan monkeys, like other species of macaques, give birth to their first babies when they are about five years old. After their first experience of childbirth, they finally become adults.

In the next chapter, I will focus on the lives of female Tibetan monkeys.

Chapter 7
Alliances among Females

1. The success story of the female, Beihen

The conspicuous social behavior of the males overshadows the females. However, the social behavior of the females, too, is very interesting. In this chapter, I would like to focus on the lives of the females, concentrating on their agonistic social interactions and their formation of alliances during the period when a rank reversal occurred among the females.

A significant upheaval occurred in the Yulingkeng group in December 1990. There was a series of violent fights among the females. This drastically changed their dominance ranks and other social relationships. A female, Beihen, who was born in 1986 and was four years old at this time, was deeply involved in this change in the social relationships among the females before and after the upheaval.

My first encounter with Beihen was in December 1989, when she was three years old. Beihen was given her name because of a big scar on her back when she was little. She was a small and quiet monkey. Her mother had died in 1988. To be honest, I was very fond of Beihen, and I was mentally rooting for her. However, she was so weak that she could not enter the central area of the provisioning site to eat the corn that had been scattered there. Yet, Beihen, who was playing far away from the site, did not seem to regret this. Despite being a low-ranking female, Beihen was carefree and spent her childhood quite normally.

When I resumed my observations in November 1990, Beihen had matured to some extent. Since she was still as shy as a child, she would screen her buttocks with her hands whenever the males attempted to mate with her.

At this point, Baijamao and Zhi had held the first and second ranks among the females since 1989.

Photo. 7–1 A coalition formed during an agonistic interaction.

It is difficult to determine whether a linear hierarchy that is actually functional exists among female Tibetan monkeys in the wild. However, on some occasions, I observed dominant-subordinate relationships among the females in the Yulingkeng group in the woods as well as at the provisioning site. For example, if an individual was eating good food, she would sometimes leave at the approach of another individual. In this case, the latter individual is dominant over (has a higher rank than) the former. In other words, dominance among the females entails having a prerogative over the food resources. Here, we will not focus on whether or not the higher-ranking females can produce more offspring; rather, we will confirm that they possess attributes that can elicit supplanting or retreating behavior from the other females.

The outcome of aggressive interactions usually depends on the dominance relationships between the individuals involved in these interactions. However, I observed that in actual fights, the subordinate individual would occasionally display aggressive behavior toward the dominant one, who would sometimes run away. In addition, a low-ranking female in a consortship with the alpha male would occasionally become aggressive and begin a fight with dominant females (Photo. 7–1).

At the beginning of my study in November 1990, Beihen appeared to be in this position. She had entered into a consortship with Duyanlong, who had become the alpha male. He was very robust and completely different from what he was in the previous year. Although Duyanlong protected Beihen whenever she was involved in a fight, the other females tried to attack her collectively. Beihen was therefore in a very unstable position.

Observational case: Beihen was almost attacked

At 1:10 p.m. on November 24th, all the members of the group arrived at the provisioning site. Teng, Zhi's daughter, screamed at Beihen, who was sitting beside Duyanlong. While Duyanlong bit Teng, Zhi chased Beihen. Duyanlong, in turn, chased Zhi and threatened her. However, Zhi continued threatening Beihen even though Duyanlong was chasing her. Baijamao and other females also threatened Beihen from a little distance away. Beihen was scared and huddled up with fear. The previous alpha male, Huangmao, who was now the second-ranking male, chased Baijamao. When he caught up with her, however, he mated with her. This behavior was typical of Huangmao, even though he was now the second-ranking male. I wondered if he was really trying to protect Beihen. Duyanlong threatened and grabbed one of the females, Feimao. Beihen rallied behind Duyanlong and displayed an amazing behavior; that is, she simultaneously performed the presenting behavior to the high-ranking females, Zhi and Baijamao, while threatening Feimao. The females continued screaming at Beihen. Baijamao and Ye put their arms around each other as if to affirm their alliance.

If Duyanlong had not been present, the females might have collectively attacked Beihen. The alliance among these females appeared to be very strong. However, on the following day, November 25th, Baijamao and Zhi suddenly began supporting Beihen.

If an individual A was the enemy of another individual Z, and a third individual B attempted to attack Z, I will summarize the situation as "B supported A." If A and B together attacked Z, I will summarize the situation as "A and B formed a coalition." If A and B regularly formed a coalition in agonistic interactions, I will summarize the situation as "A and B formed an alliance."

Observational case: Beihen suddenly became a high-ranking female

At 2:40 p.m. on November 25th, Beihen threatened Feimao while they were in the woods. Zhi, too, threatened Feimao, and Beihen then solicited the support of Zhi, who formed an alliance with her. Beihen hugged Zhi from behind, putting her arms around Zhi as if to threaten the common enemy while affirming her alliance with Zhi. Feimao avoided them and left.

At 3:00 p.m., Baijamao threatened Feimao. Immediately, Duyanlong approached and mated with Feimao. I thought, "You act just like Huangmao, Duyanlong." Beihen approached Feimao, inspected her genitals, and mounted her. Beihen then returned to Baijamao and Teng and threatened Feimao again.

On the following day, November 26th, Baijamao died. After her first fall on my first observational day in 1989, Baijamao fell from the cliff many times. Although she resembled a witch with the many scars on her face, she was still the highest-ranking female. However, on this day, she was found dead in the basin of a waterfall. Since she was found in the water, I surmised that she might have drowned after being knocked unconscious as a result of falling from the cliff. However, after picking up her carcass, I noticed that her head was split in half. I am not of the opinion that Beihen pushed her off the cliff in order to secure a higher rank. Nevertheless, as a result of this incident, the group lost its highest-ranking female, and Zhi lost her ally. Beihen behaved aggressively even in the absence of direct support from Duyanlong, and came to occupy a higher rank among the females.

Beihen, Zhi, and Teng later became dominant over the other females. On November 30th, Beihen, Zhi, and Canglian appeared to have been bitten by some males—they had distinct scars on their bodies. However, Beihen, Zhi, and Teng retained their high ranks.

Beihen, who did not have a mother or sisters and was only four years old at that time, came to occupy a position similar to that of an alpha female. This came as a surprise to me because, as I will discuss later, such a situation is extremely rare among the Japanese monkeys in Arashiyama, who maintain their dominance ranks by means of alliances among related females.

Further, another upheaval occurred on December 8th. Ye, who was a low-ranking female in the group, defeated the high-ranking female, Zhi, in an intensely aggressive fight.

Observational case: Ye was provoked

At 1:34 p.m. on December 8th, a two-year-old female—Baijamao's daughter, Sanjiaolian—got angry and screamed at Teng while they were in the woods. Teng retaliated by threatening Sanjiaolian. These females antagonized each other. The adult male, Erxiong, supported Sanjiaolian. After Baijamao's death, Erxiong protected Sanjiaolian in aggressive interactions. However, Zhi and Beihen, who had formed an alliance, both supported Teng, and together, they threatened Sanjiaolian. Wushang decided to support Sanjiaolian and bit Teng, who was still very angry. Surprisingly, she jumped on Erxiong, who of course, fought back and bit her. Wushang, Erxiong, Zhi, Beihen, and Teng grappled with each other. Beihen and Zhi, who were being chased by Wushang, fell into the basin of a waterfall.

After some time, Beihen and Zhi emerged from the waterfall soaking wet. Teng and Sanjiaolian had stopped fighting. Zhi and Beihen hugged each other tightly. I thought "Everything is over now."

However, Beihen suddenly screamed at and grabbed Zhi. Duyanlong, too, ran to them and grabbed Zhi, who ran away. During this interaction, Ye continued screaming at Zhi and Beihen from a short distance away. Ye was standing behind Duyanlong. Ye had been in a consortship with Duyanlong for two days—November 28th and 29th—after which she often lingered beside Duyanlong. At one point, Beihen appeared to have begun a fight with Zhi; however, they hugged each other again and began screaming at Ye. Ye continued screaming at them from behind Duyanlong, who then mated with her. Teng approached Beihen and Zhi, but Duyanlong drove her away. However, Beihen, Zhi, and Teng continued to threaten Ye, who was slowly driven into a corner. Duyanlong followed them and mated with Teng, immediately after which he bit her again.

Hua then arrived at the spot. Beihen mounted her. Hua, too, threatened Ye. However, Beihen suddenly bit Hua, and there was chaos again. Duyanlong bit Zhi, and Duyanlong, Zhi, Beihen, Hua, and Ye huddled up and rolled down the hill.

In summary, Zhi, Beihen, and Teng were trying to attack Ye, while Duyanlong was trying to protect her. Hua, who somehow became involved in this fight, appeared to be on Beihen's side. Beihen, however, bit Hua. She may have wanted to assert her dominance over Hua while soliciting her support. Hua, however, ultimately opposed her. Zhi, Beihen, Hua, and Ye fell into the basin of the waterfall again.

Finally, a fierce battle took place between the Zhi Beihen pair and the Ye Hua pair, in which they attempted to bite each other. Duyanlong and Huangmao attempted to put an end to this fight because of its intensity and fierceness. However, the females did not stop their battle and continued biting each other's bodies or fingers. Fights between males end quickly because they can tear each other's bodies with their sharp canines. Fights between females, however, may continue for a long time.

At 2:03 p.m., Zhi, who had been bitten by Ye, finally stopped struggling. After an intense battle that had continued for 30 minutes, the impromptu team of Ye and Hua defeated the alliance of the higher-ranking females Zhi, Teng, and Beihen.

Ye, who was usually quiet, continued chasing Teng throughout that afternoon with a devilish expression on her face.

The following day—December 9th—Zhi, who had lost the fight, did not arrive at the provisioning site. I looked for her, but the monkeys knew where she was better than I did. I followed Feimao, who had left the group, and I found Zhi resting quietly in the woods. Feimao was grooming her. In a few days, Zhi's physical condition improved. However, after the incident that occurred on December 8th, Zhi became weak and was continually afraid of Ye. Further, Zhi's daughter Teng became the lowest-ranking female. Since Zhi threatened Teng and redirected any aggression toward Teng whenever she was about to be attacked, the incident that occurred on December 8th was a huge misfortune for Teng. Did Ye and Hua secure high ranks because of their victory on December 8th? What happened to Beihen? Beihen formed a new, strong alliance with Hua and maintained her high-ranking position. One week later, after the relationships among the females had stabilized to some extent, I often observed the following situation.

Observational case: Beihen maintained her high-ranking position

At 10:00 p.m. on December 18th, all the members of the group arrived at the provisioning site to eat the scattered corn. Hua approached Teng and took hold of her buttocks as if to mount her. Hua groomed Teng for a short time, holding her buttocks down. After she had been groomed, Teng left Hua and tried to maintain some distance from her. However, Hua still followed her. Beihen arrived, approached Hua, hugged her, and stared and screamed at Teng.

During this period, Duyanlong did not support Beihen in all the fights among the females. Yet, the memory that Duyanlong had supported her contributed significantly to the maintenance of Beihen's social position. Although she occasionally displayed sudden aggressive behavior in agonistic interactions, she was usually willing to groom the subordinate females. Beihen appeared envious of Zhi and Canglian, who were holding their infants; hence, she frequently approached them in order to bridge and often groomed their infants.

Did Beihen establish her high-ranking position herself?

On December 15th, Duyanlong formed a consortship with Feimao. He was then in a consortship with Hua on the 20th, with Ye on the 23rd, and finally, with Zhi on the 25th. Duyanlong was fond of everyone. I observed the following situation on January 8th.

An observational case: Duyanlong now turns his attention to Zhi

At 8:30 a.m. on January 8th, I heard a noise from the woods on the opposite bank. I assumed that this was the sound of Beihen screaming at Zhi. When I arrived at the spot, I saw—as I had expected—Beihen, Zhi, and Duyanlong.

Duyanlong bit Beihen, despite which she grabbed Zhi's body. Zhi bit Beihen in return. These three monkeys grappled with each other and rolled down the cliff. Duyanlong chased Beihen, who ran away.

Hua arrived at the spot and approached Zhi in order to inspect her genitalia. Ye approached and chased Zhi; she still attacked Zhi alone. Duyanlong then mated with Zhi. Hua and Beihen hugged each other, and the other females, too, approached and threatened Zhi. Although Zhi tried to avoid these females and leave, they followed her. Duyanlong, too, followed them; Beihen, however, remained at the spot and began eating.

Thus, although Beihen was still dominant over Zhi because of her alliance with Hua, Duyanlong lent his support, which Beihen had counted on, to Zhi. Beihen was in an unstable position again. This situation continued until I interrupted my observations on January 11th.

My next visit to Huangshan was on March 25th. I had been worried about Beihen because I assumed she would have fallen in rank again on account of the incidents that occurred in January. However, even if this were to be the case, Beihen may not have had any regrets, and I expected her to frolic in the woods just as she had done when she was three years old. However, Beihen was all right —she had maintained her rank. Whenever the females fought, Beihen approached after the fight and mounted Teng and hugged Hua. I assumed that Duyanlong's support was effective in enabling Beihen to secure a high-ranking position in the group; however, when his support was directed toward Zhi, the mating season, too, had ended, and the support of the males in a consortship became less effective than the alliances among the females.

2. Social behavior of the females in agonistic interactions

As described in some observational cases, the females hugged each other, performed genital inspections, and as though they were mounting, one female would hold the other's buttock. Table 7–1 summarizes the social context in which these behaviors occurred. These social behaviors occurred more frequently when the dominance ranks of the females were destabilized by upheavals in the group. After having described male social behavior in Chapter 4, I will now introduce each type of social behavior among the females.

Genital inspection

A female Tibetan monkey often looks at and touches the genitalia of her subordinate females. The females also perform behavior similar to that in which

Table 7–1 Social behavior of the females in each context.

Previous interactions	A's behavior toward B					
	Hugging		Hold bottom		Genital inspection	
	Number of instances	(%)	Number of instances	(%)	Number of instances	(%)
Aggressive interactions						
A and B attacked another individual	47	(22.0)	9	(5.1)	1	(0.4)
B attacked another individual	31	(14.5)	30	(17.1)	5	(19)
B was attacked by another individual	7	(3.3)	14	(8.0)	17	(6.5)
A attacked B	8	(3.7)	13	(7.4)	35	(13.4)
Other aggressive interactions	18	(8.4)	15	(8.6)	9	(3.4)
Hugging, hold bottom, and genital inspection	24	(11.2)	5	(2.9)	13	(5.0)
Others	79	(36.9)	89	(50.9)	182	(69.5)
Total	214	(100)	175	(100)	262	(100)

the males attempt to monitor the sexual cycles of the females. It would be interesting if the females do, in fact, monitor the sexual cycles of other females. By this behavior, the dominant female may confirm her dominant-subordinate relationship with the subordinate females. The females often displayed this behavior when their ranks changed, although the females sometimes performed this behavior when their dominance ranks were stable.

Male Tibetan macaques do not engage in ritualistic social behavior that is directed solely by the dominant individuals toward the subordinate ones; however, the latter frequently perform the presenting and penis-showing behaviors in front of the dominant individuals. It is reported that the males of some primate species confirm their dominance relationships by performing the mounting behavior. However, the results of my study do not indicate that the dominant males mount subordinate males more frequently than vice-versa. It is interesting that while the dominant females engage in behavior that affirms their dominant-subordinate relationships, the dominant males do not.

The females frequently performed genital inspections after one individual A attacked another individual Z. Individual A would then follow Z, hold up her buttocks, and inspect her genitalia. It appeared as though the females performed genital inspections in order to doubly confirm their dominance relationships.

Hugging

Female monkeys hug while facing each other. Unlike the males, who embrace each other while swaying their bodies, the females hug each other tightly, putting their arms around each other (Photo. 7–2). When two females collectively attacked another individual, they frequently hugged each other as if to affirm their alliance. When two females are faced with a common enemy, one female sometimes puts her arm on the other's shoulder, and they threaten the enemy as though they are seeking support (Photo. 7–3).

Hold bottom

A female monkey sometimes holds the buttocks of another female (Photo. 7–4). I did not observe this behavior on even one occasion before the rankings were overturned in 1990. Beihen alone mounted the other females completely, holding their hind limbs with her own, in a manner similar to that of the males. Females perform this behavior when one female and another female collectively attack the third female, or when another female attacks the third female. This behavior seems to imply "I support you, but I am still dominant over you."

Photo. 7–2 Females hugging each other.

Photo. 7–3 One female hugging another and threatening a third.

Photo. 7–4 "Hold bottom" behavior.

3. Formation of allianccs among the females

Table 7–2 and Fig. 7–1 summarize the change in the social relationships among the females between November 1990 and January 1991. During period 1 in Table 7–2, Baijamao and Zhi held the first and the second rankings, respectively, and Beihen had formed a consortship with the alpha male, Duyanlong. During period 2, Duyanlong continued protecting Beihen from the other females. Once the high-ranking females, Baijamao and Zhi, supported Beihen, Beihen began to behave aggressively toward the other females. On November 26th, after Baijamao had died of her fall from the cliff, Zhi and Beihen formed an alliance and behaved aggressively toward the other females. However, on December 8th, during period 3–1, there was a fierce battle between the alliance of Zhi and Beihen and that of Ye and Hua. Zhi and Beihen, who had held the higher ranks in the group, lost the battle. However, Ye and Hua did not secure higher ranks. During period 3–2, Beihen and Hua formed a new alliance and maintained their high ranks. After December 25th, during period 3–3, Duyanlong entered into a consortship with Zhi and supported her. However, Beihen, who had formed an alliance with Hua, opposed Zhi. When I returned to Huangshan on March 25th, after the interruption of my observations between January 11th and March 24, 1991, Beihen and Hua still maintained their high ranks. Under these circumstances, I completed my observations on April 20th.

Table 7–2 Consortships, female alliances, and female ranks during the rank reversal.

Study Period	1 Nov. 18, 1990– Nov. 24, 1990 7 days	2 Nov. 25, 1990– Dec. 7, 1990 13 days	3–1 Dec. 8, 1990– Dec. 10, 1990 3 days	3–2 Dec. 11, 1990– Dec. 24, 1990 14 days	3–3 Dec. 25, 1990– Jan. 11, 1991 18 days
Consorts	Duyanlong-Beihen (7 days) Huangmao-Zhi (7 days)	Duyanlong-Beihen (11 days) Duyanlong-Ye (2 days) Huangmao-Zhi (13 days) Wushang-Ye (6 days)	Erxiong-Beihen (3 days)	Duyanlong-Feimao (5 days) Duyanlong-Hua (2 days) Duyanlong-Ye (2 days) Huangmao-Hua (14 days) Wushang-Ye (12 days) Erxiong-Beihen (14 days)	Duyanlong-Zhi (18 days) Huangmao-Canglian (14 days) Erxiong-Beihen (18 days)
Alliances	Baijamao-Zhi	Beihen-Zhi	Beihen-Zhi Ye-Hua	Beihen-Hua	Beihen-Hua
Order of Ranks	Baijamao Zhi Teng Hua Ye Feimao Canglian Tougou Beihen	Beihen Baijamao* Zhi Teng Hua Ye Feimao Canglian Tougou	?	Beihen Hua Feimao Ye Canglian Tougou Zhi Teng	Beihen Hua Feimao Ye Canglian Tougou Zhi Teng

*: Baijamao died on Nov. 26, 1990.

Fig. 7–1 Rank reversals among the females.

This change in the ranks of the female Tibetan monkeys may be easily understood if we assume that they were playing an alliance formation game in which each female attempted to form the best possible alliance on the basis of each female's strength in agonistic interactions, while taking into account the support of the high-ranking males.

Primatology includes a concept known as "depending ranks," that is, a subordinate individual can become "dominant" by securing the support of a third individual. In contrast, an individual's "basic rank" is determined in situations that involve only the individual and his or her opponent and not a third party. However, since Tibetan monkeys live in groups, they rarely encounter situations involving only two individuals. I am inclined to believe that their ranks are essentially depending ranks and they always attempt to manipulate their behavior in situations involving other members of the group. Basic ranks may merely represent particular situations in their minds, for example, "Since I have the support of a male, it is acceptable for a certain individual to form a coalition with another under these circumstances" and "In this situation, it will be advantageous for me to support a certain individual, even if the others form a coalition with someone else."

The alpha male, Duyanlong, began to attack Beihen instead of attacking Zhi, as presented in the upper part of Fig. 7–2. The female that he supported in agonistic interactions also changed from Beihen to Zhi, as presented in the lower part of Fig. 7–2. The relative power of each female in an agonistic interaction was changeable, depending on Duyanlong's support. This balance of power affected the females' choices of whom they ought to support or form an alliance with in order to secure the highest possible rank. Their choices resulted in the change of alliances presented in the upper part of Fig. 7–3. The females who frequently formed alliances in agonistic interactions frequently hugged each other, as presented in the lower part of Fig. 7–3. As a result, the females who formed successful alliances secured higher ranks.

Baijamao and Zhi; Zhi and Beihen; and Beihen and Hua were the pairs of females who formed successful alliances. These were not mother-daughter or sister pairs.

Were these females close friends who frequently groomed each other as their alliance partners?

Grooming is not merely an indication of the familiarity between two individuals; rather, it functions as a type of currency among primates. This is because grooming is an altruistic behavior that requires the individual who performs it to spare his or her time, even if not a lot of physical energy. The individuals performing the grooming reduce their own fitness while enhancing the fitness of the recipients of the grooming. If the recipient is an unrelated individual, then grooming is a mere waste of energy and time for the individuals performing it unless their action is reciprocated by the recipient. Yasuyuki Muroyama analyzed the initiation of grooming interaction, the change of the roles in grooming, and solicitation of grooming among Japanese monkeys and found that an individual who groomed another more frequently solicited a return of the favor by lying down or crying before the recipient, than did the individual who was groomed.(3) It appeared as though the individual who had performed the grooming was saying, "Please groom me this time. It's your turn now." While Japanese monkeys who were related did not expect any recompense, such expectations did exist between unrelated individuals. Therefore, the recipient of the grooming had to reciprocate the action to the individual who had performed the behavior.

Unlike Japanese monkeys, Tibetan monkeys did not cry in order to solicit grooming in return.(4) A male sometimes groomed a female for a very short period and then received a longer grooming in return. This grooming itself appeared to be performed in order to solicit grooming in return.

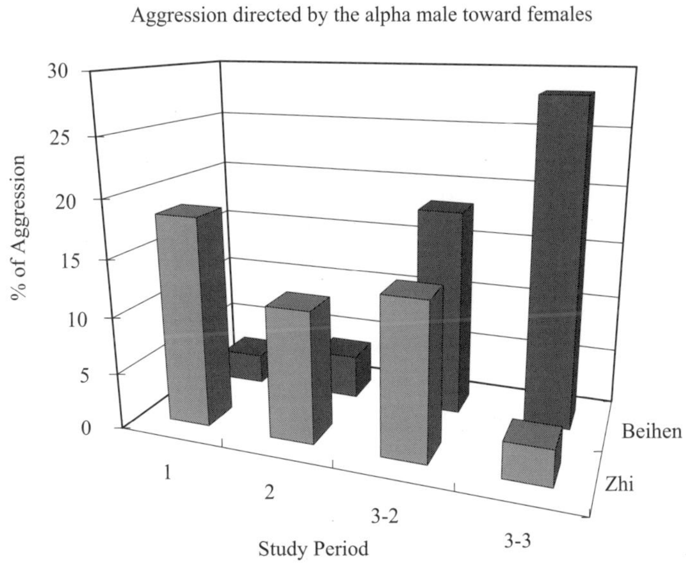

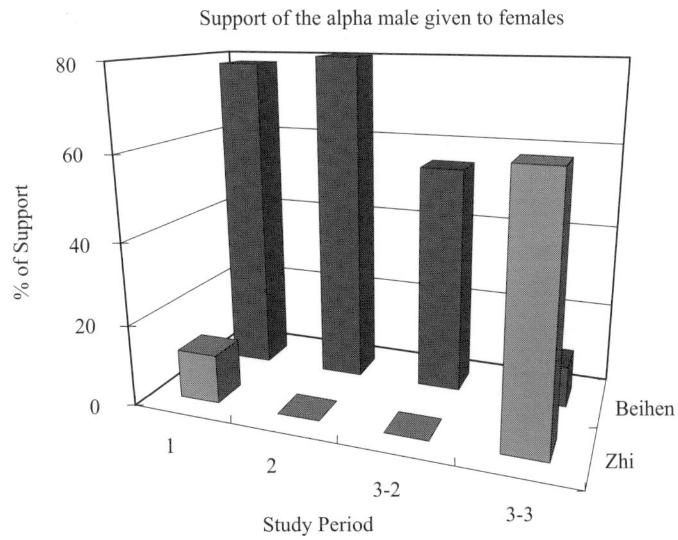

Fig. 7–2 Changes in male aggression and male support given to females. Pecentage of aggression: (Number of occurances of aggression toward the females, Beihen or Zhi) / (Total number of occurances of aggession towards females). Pecentage of support: (Number of times support was provided to the females, Beihen or Zhi) / (Total number of times support was provided to females). Support: The alpha male attempted to attack the enemy of Beihen or Zhi, when Beihen or Zhi were involved in aggressive interactions.
Study Period: 1 = Nov. 18, 1990 – Nov. 24, 1990; 2 = Nov. 25, 1990–Dec. 7, 1990; 3–2 = Dec. 11, 1990–Dec. 24, 1990; 3–3 Dec. 11, 1990–Dec. 24, 1990.

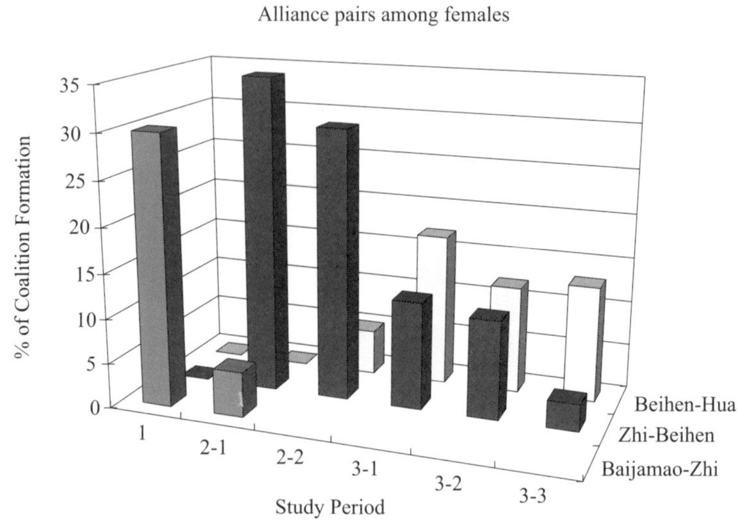

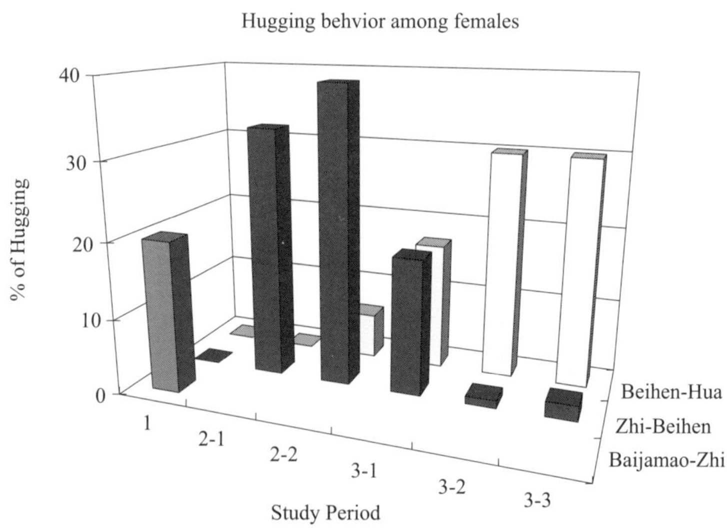

Fig. 7–3 Changes in the alliances and frequency of hugging among females.
Percentage of coalition formation: (Number of times a coalition was formed between the female pairs, Baijamao-Zhi, Zhi-Beihen, or Beihen-Hua) / (Total number of times a coalition was formed between females).
Percentage of Hugging: (Number of times hugging occurred between the female pairs, Baijamao-Zhi, Zhi-Beihen, or Beihen-Hua) / (Total number of times hugging occurred between females).
Study Period: 1 = Nov. 18, 1990–Nov. 24, 1990; 2 = Nov. 25, 1990–Dec. 7, 1990; 3–2 = Dec. 11, 1990–Dec. 24, 1990; 3–3 Dec. 11, 1990–Dec. 24, 1990.

We still do not know how Tibetan monkeys solicit a reciprocation of this altruistic behavior. There are some reports, however, that among other primates, the recompense assumes forms other than grooming. For example, Vervet monkeys supported the individual who had groomed them in agonistic interactions involving that individual more frequently than they did when they had not received grooming.

I wondered whether this was the case among Tibetan monkeys as well; that is, whether the females who often supported other females in agonistic interactions were those who owed a favor to those individuals. However, the female Tibetan monkeys who formed alliances and secured higher ranks in the Yulingkeng group in 1990 were not necessarily those who performed or received grooming. Female Tibetan monkeys may choose allies who benefit them strictly on the basis of the relative power balance among the females; this is more important than the performance and reciprocation of grooming.

In complex interactions involving multiple individuals in such varying circumstances, Tibetan monkeys must require social knowledge that is much more complex than that required in interactions involving only two individuals. In such a social environment, the monkeys must understand the relative power of the individuals in agonistic interactions that involve themselves and others and the balance of power.

Primates, especially baboons and macaques, often form coalitions in agonistic interactions. For example, among yellow baboons (*Papio cynocephalus*), two or more low-ranking males approached a high-ranking male who was accompanied by a female, and separate the pair in order to get to the female. A male who was likely to be a suitable ally was chosen as a partner in the coalition. Once this coalition was successful, this popular male often secured a female.(6) Thus, baboons chose partners with whom they are likely to form a successful coalition by evaluating the relative power of each individual in fights, which could influence the outcome of the agonistic interaction.

F. de Waal meticulously observed the process by which captive chimpanzees competed for higher ranks.(7) The alpha male could usually monopolize the females and mate with them. Within the group, however, the alpha male needed the support of the third-ranking male to ensure that his position would not be usurped by the strong, second-ranking male. The second-ranking male also needed the support of another male, the third-ranking male, to enable him to better compete with the alpha male. Each male understood the balance of power, namely, that the support of the third-ranking male was crucial. Therefore, the third-ranking male took advantage of the situation and was able to mate with more females than was usually the case. Similar interactions were also observed among wild chimpanzees.(8)

Table 7–3 Changes in the dominance ranks of the females.

Rank	Study Period					
	Dec. 1989 – Apr. 1990	Dec. 1990 – Dec. 1990	Dec. 1990 – Apr. 1990	Dec. 1991 – Apr. 1992	Sept. 1992 – Nov. 1992	Apr. 1993 – Apr. 1993
1	Baijamao	Beihen	Beihen	Hua	Hua	Hua
2	Zhi	Zhi	Hua	Ye	Ye	Ye
3	Teng	Teng	Feimao	Canglian	Canglian	Canglian
4	Hua	Hua	Ye	Tougou	Tougou	Tougou
5	Ye	Feimao	Canglian	Eshang	Beihen	Eshang
6	Feimao	Canglian	Zhi	Beihen	Eshang	Beihen
7	Canglian	Tougou	Teng	Feimao	Feimao	Feimao
8				Zhi	Zhi	Zhi
9				Teng	Teng	Teng
10					Sanjiaolian	Sanjiaolian
11					Keai	Keai
	(Beihen)	(Eshang)	(Eshang)	(Sanjiaolian)	(Baixiang)	(Baixiang)
	(Tougou)	(Sanjiaolian)	(Sanjiaolian)	(Keai)		
	(Eshang)	(Keai)	(Keai)	(Baixiang)		

The dominance ranks of females who were 4 years old and older are listed. The names of females who were 3 years old and younger are enclosed in parentheses.

As described in Chapter 4, primates can understand the relationships among the members of their group. They are able to step out of their self-centered perspectives and understand the social relationships that other individuals are involved in. In addition, primates may be able to understand the relative power of each member of the group in agonistic interactions and the balance of power in these relationships.

4. Social relationships among unrelated females

When I arrived at Huangshan in December 1991, Beihen was still behaving aggressively; however, her rank had fallen considerably (Table 7–3). Beihen and Hua were both high-ranking females in April 1991, but later, only Beihen became a low-ranking female. Zhi, whose rank was expected to rise again because of her consortship with Duyanlong, ultimately remained a low-ranking female.

As has already been described, the ranks of the females changed significantly within the span of a few weeks, during the study period from December 1990 to January 1991. This may have been an extreme case since there was no similar rank reversal after December 1991. However, Table 7–3

Photo. 7–5 Mothers resting together.

shows that the ranks of the females changed every year between 1989 and 1992. A characteristic feature of Tibetan monkeys is that daughters do not necessarily inherit their mothers' ranks.

Beihen gave birth in the spring of 1991. The female who was closest to her was Tougou. Tougou was the same age as Beihen and she, too, had her first baby that year. I will discuss Beihen's life later.

Observational case: Tougou became a new friend to Beihen, who was now low-ranking

At 3:48 p.m. on December 20, 1991, Tougou approached Beihen, who was holding her first baby, Beixiao, and sat down facing her. Tougou lifted Beixiao's legs up to her head, and Beihen and Tougou licked Beixiao's genital area; that is, they engaged in bridging. Beihen then groomed Beixiao for a short time. Tougou's baby, Toutai, approached Tougou and hid in her belly. Tougou lifted up Beixiao again and began bridging. Beihen and Tougou groomed Beixiao, holding her between them. Further, Tougou groomed Beixiao, Beihen groomed herself, and Toutai groomed Beihen.

Tougou then stood up and left Beihen. She encountered Duyanlong on her way, and attempted to pass by him by merely performing the presenting behavior. However, Duyanlong followed Tougou. Beihen, too, stood up and followed them; Beixiao followed Beihen.

Tougou was in a consortship with Duyanlong during the winter of 1991. Duyanlong no longer protected Beihen, but defended Tougou in the case of agonistic interactions among the females. Although we humans may assume that Beihen was not happy about the relationship between Duyanlong and Tougou, Tougou was Beihen's closest friend at that time.

Tougou, who was Ye's daughter, would spend time with her mother in the woods in 1990. After Tougou had a baby, however, she stopped associating with Ye to the same extent. I observed that other young adult females also behaved in the same manner. Related females, especially mothers and their daughters aged four years old or younger, associated with and often groomed each other. However, when the daughter became an adult and gave birth to her own baby, her relationship with her mother weakened. The females who became mothers during the same year frequently came together, groomed each other, and bridged. Fig. 7-4 presents Beihen's grooming partners from the late fall in the years 1989, 1990, and 1991, to the following spring. Beihen changed her grooming partners every year. Another adult female, Zhi, who had two adult daughters, Teng and Canglian, also had unrelated grooming partners as well and changed her grooming partners every year.

On reading about the social relationships among the female Tibetan monkeys, as described in this chapter, the reader may receive the impression that the genus *Macaca* includes some strange species.

Research on the society and social behavior of primates first began with a study on Japanese monkeys belonging to the provisioned groups on Koushima Island and in Takasakiyama, and was then extended to wild populations, such as those on Yakushima Island. Due to this background, many Japanese people have the following impression of these provisioned Japanese monkeys: there is a strict dominance rank hierarchy among the individuals, their daily interactions are restricted to those between related females, and related females form coalitions to fight those who are not related to them. Since strong mothers unfailingly support their daughters, a daughter whose mother is dominant over a certain female becomes dominant over the daughter of that female. Daughters inherit their mothers' ranks, and generally, a mother is dominant over all her daughters. Some readers may be aware of another rule that females are dominant over their elder sisters. This rule does not hold true for the small group of wild Japanese monkeys on Yakushima Island. Nevertheless, their dominance ranks are based on their matrilineal relations, and related high-ranking females tend to support each other and maintain their ranks.

While several studies have been conducted on Japanese monkeys, others have focused on other macaque species. The studies on Japanese monkeys and rhesus monkeys, however, are the ones that have produced the largest amount of

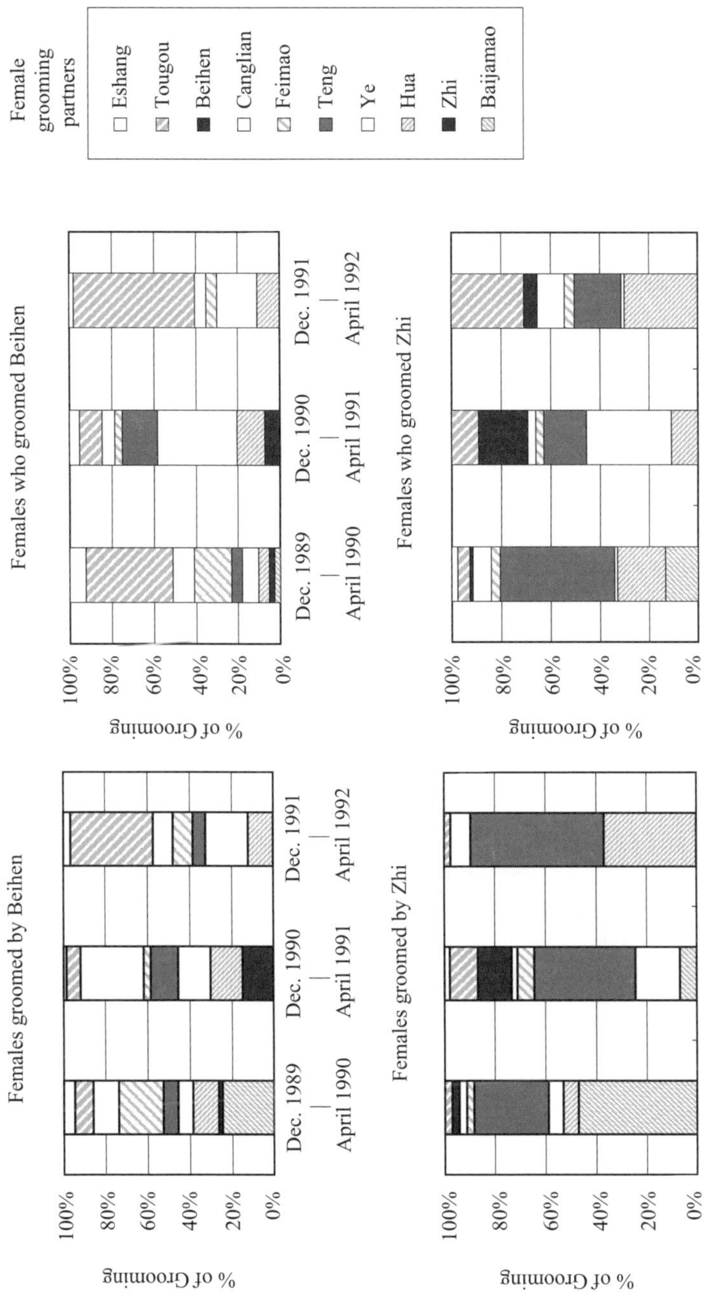

Fig. 7–4 Changes in grooming partners among the females.

research material. It is because of this historical background that references to macaques tend to call up images of the society of Japanese monkeys.

However, the extensive studies conducted on other macaques, especially during the 1980s, have led to the realization that it is not only Tibetan monkeys but other monkeys, too, whose social relationships are quite different from those of Japanese monkeys. A characteristic that is common to most macaques is that they live in multi-male, multi-female groups, although they sometimes form one-male groups. In addition, they basically form groups based on female philopatry and male dispersal, and have a linear dominance hierarchy. However, although they share a common basic social structure, the social relationships and social behaviors differ among species. Moreover, each species belonging to the genus *Macaca* can be classified into the following two categories.(9) The first category includes Japanese monkeys and rhesus monkeys; these monkeys are referred to as "nepotistic," and they support individuals who are related to them. They have a "despotic" dominance style. The monkeys in the second category are referred to as "egalitarian"; they are not very concerned about kinship and have a "relaxed" dominance style. In this category, related females are not very close and the dominance ranks in the group are not strictly determined for a long period of time. This category includes Tibetan monkeys, Barbary monkeys, stump-tailed monkeys, Bonnet monkeys, and Moor monkeys (*Macaca maurus*). However, detailed observations have not been conducted on the wild populations of some species.

Some believe that the relationships between monkeys are of the latter category—egalitarian, and with a relaxed dominance style—because the high-ranking individuals cannot be exceedingly aggressive toward the low-ranking individuals since the group must be able to compete in the case of intense rivalry with other groups.(10) On the other hand, Shuichi Matsumura and Takashi Kobayshi developed a model that allowed both nepotistic and egalitarian species can be evolved.(11) This model implies that both options allow for stability if all the members of the group prefer a nepotistic or an egalitarian lifestyle. Further, the individuals that adopt opposite lifestyles cannot intrude into each other's groups because they have a higher fitness than those who adopt the opposite lifestyle.

Why have Tibetan monkeys formed the latter type of society? On the basis of my observations thus far, I can identify the following characteristics in the social relationships among female Tibetan monkeys. First, the females do not form close social bonds with related females in their daily lives, and they often have unrelated grooming partners. For example, females who have a baby in the same year frequently groom each other. The females do not necessarily support their relatives in agonistic interactions. Their dominance ranks are relatively ambiguous and change quite often, and daughters will not inherit their mothers' dominance ranks. In this type of society, are the dominance ranks determined by each individual's ability to fight in agonistic interactions? The answer to this is "No." The alliances among unrelated females exert a greater influence on the dominance ranks of Tibetan monkeys than does the strength of each individual. The females who have the ability to understand the balance of power and make good choices in the social interactions involving three or more individuals form successful alliances. Unlike in the case of Japanese monkeys, the interactions with unrelated members of the group play a more important role in the social life of Tibetan monkeys than the interactions with the related members of the group. Since I was unable to identify the relationships between the social groups during my short visit at Emeishan, I am uncertain whether high-ranking female Tibetan monkeys cannot be very aggressive toward low-ranking members of the group because of the competition with other groups.

In addition to the characteristics I have mentioned so far, other factors may also be related to this issue; the close proximity between individuals, the fact that they do not have soliciting behavior for grooming, and the fact that there is little conflict between the mother and the baby during the weaning period. The degree of dominance displayed by the males toward the females on the basis of sexual dimorphism and the way in which the males interfere in agonistic interactions among the females may be also related. There may have been certain historical factors that allowed the social groups of the Tibetan monkeys to evolve in the environment of Huangshan. All these factors are interrelated.

In this chapter, I have focused on Beihen's experiences, but I have not yet summarized the data on female Tibetan monkeys, which is comparable with that on other macaque species. In the future, however, as a researcher who has observed Tibetan monkeys for several years, I would eventually like to explore the evolution of macaques and their societies.

Chapter 8

Infanticide

1. Beihen's baby was killed

Another incident occurred during the spring of 1993—a possible infanticide took place.

Beihen gave birth to a female baby on April 20, 1993. This was her second baby; the first, Beixiao, was born in 1991. The incident occurred just while I was trying to think of a name for the newborn baby.

Observational case: Beihen's daughter was killed

On April 24, 1993, we were climbing the mountain in order to observe the monkeys of the Yulingkeng group. At 8:10 a.m., when we had almost arrived at the valley where all the monkeys were assembled, we heard them making loud noises. Ten minutes later, we found the monkeys in the woods and began our observations. At that moment, Dr. Wada noticed that the female baby whom Beihen was holding was dead. We noticed injuries on her stomach and chest that resembled the cuts of a sharp knife. Beihen, who remained beside her baby, appeared to be at a loss. Since our assistant informed us that Beihen had come to the provisioning site with her baby when the monkeys were fed corn at 7:00 a.m., I suspected that the baby had been injured and had died after the commotion we had heard that morning. Sanxiong, Duyanlong, and Hua were sitting around Beihen, and there was blood smeared on Sanxiong's mouth. Apart from the dead baby, no other monkey was injured. Beihen picked up her dead baby and carried her into the woods.

Observational case: Canglian's daughter also died

On April 18th, six days before the death of Beihen's daughter, another baby who had been born just this spring also died. This was Canglian's daughter.

When we began our observations on April 11th, this baby had already been born —according to the people of the village of Fuxi, she had been born on April 4th. The first time I saw this baby, she had a sharp cut on the joint of her left hind limb. She could not hold on to her mother's belly well and would swing her injured limb. The cut gradually widened because of the weight of her limb. This may also have been due to the fact that she was used in bridging between her mother and the other females. The result of this behavior was ironic since bridging between the females appears to be a celebration of the birth of a baby. This baby finally died on April 18th.

Observational case: Eshang's baby also died

Although I did not witness this event, Ehong, too, died after I had completed my observations and left Huangshan in 1992. Ehong, Eshang's daughter, was born on April 24th. At 2:00 p.m. on May 23rd, Eshang was found holding her daughter who had sustained a serious injury. The baby's right hind limb had been cut off and her intestines were exposed. This baby died soon after the researchers found her.

I suspect that all the injuries sustained by these three babies were inflicted by the incisors of male Tibetan monkeys. Currently, there is no other species of wild animal in our study area that could possibly have inflicted this type of injury on the Tibetan monkeys. During the studies conducted between 1985 and 1993, we observed three cases of possible infanticides carried out by male Tibetan monkeys. All the dead babies were female and their mothers were all young adults. However, I am uncertain whether this was a mere coincidence or a deliberate act that could reveal a statistically significant tendency once more cases are documented.

Since Yukimaru Sugiyama's discovery of the practice of infanticide among male Hanuman langurs (*Presbytis entellus*), many cases of infanticide have been reported among other primate species and among other animals such as lions. Their implications have been discussed.(1) (2) Currently, it is held that in most cases, infanticide is one of the male reproductive strategies that allow the males to produce as many offspring as possible. In the case of a new male in a one-male group, the infants that were born before he secured the position of a resident male are not his offspring. Such a male would kill those infants in order to ensure that the females are prepared to bear his offspring as soon as possible. In a multi-male, multi-female group, the new immigrant males and low-ranking males do not have many opportunities to mate with the females. When they become high-ranking males, their position is like a new resident male in a one-male group.

I was informed that a young adult male, Gaoshan, had immigrated into the Yulingkeng group just before the death of Eshang's baby, and had secured a high-ranking position when she died. In the first and second cases, Sanxiong, who had been a low-ranking male in the Yulingkeng group, replaced Duyanlong and became the alpha male. On the basis of these facts, infanticide among Tibetan monkeys may be understood as a male reproductive strategy if these males actually killed infants. However, Sanxiong had already been the alpha male during the previous mating season, and had mated with both Beihen and Canglian. Sanxiong may thus have killed his own daughters.

Among Tibetan monkeys, the mothers are very tolerant of other females' or males' handling their babies while bridging and holding. In Chapter 6, I mentioned that the females mate with many males in order to confuse the paternity of the baby; therefore, the males know that the baby may be their offspring and do not treat it roughly. It is for this reason that the females tolerate the interactions between their babies and the males. Until I actually witnessed the death of one of the infants, I believed that this female strategy would reduce the risk of infanticide at the hands of the males. It remains risky for female Tibetan monkeys to allow others to touch their offspring. Does this mean that taking into account the risks and the costs and benefits, female Tibetan monkeys choose to tolerate the contact between their babies and the males?

As is sometimes observed in other primate species, the mothers in this group, Beihen, Canglian, and Eshang, continued holding their babies even after they had died. They spent many days grooming and licking the injuries of their dead babies. Beihen's baby was even used in bridging between Beihen and the other females after her death. It is possible that Tibetan monkeys cannot understand the concept of death or that even carcasses can be used in bridging. Nevertheless, to my eyes, this behavior appeared to be a solemn funeral for the babies.

While I believe that the above conclusion appears literary, I must reveal the true nature of the Tibetan monkeys. As the head of the dead baby lost its shape, Canglian and Beihen opened and began eating the base of its neck.

Let this be endured, as it occurred after the funeral ceremonies had entirely ended.

Chapter 9

Huddling at the Sleeping Site

1. Huddling behavior

In this chapter, I would like to discuss the social interactions that occur when the Tibetan monkeys huddle and sleep together at their sleeping site on the cliff, and analyze the composition of the huddle and the disposition of each of its members.

Sometimes—especially on very cold days—the monkeys form a huddle in order to warm up their bodies (Photo. 9–1). When they take naps on warm days, they lie on tree branches and assume the various postures they like, lying on their backs or bellies. However, during the cold weather, they assume a posture that allows them to minimize the surface area of their bodies exposed to the cold air; that is, they crouch down and wrap their hands around their bosoms. When they still feel cold, they make physical contact with each other. Two monkeys may huddle facing each other, covering each others' belly, another may place his or her belly against the backs of the first two, and many monkeys together eventually form one huddle. I call this group a "huddling cluster" or a "spit dumpling of monkeys" ("saru-dango" in Japanese). ("saru" means "monkeys" in Japanese, and "dango" refers to "round flour dumplings." Here, this metaphor is used to describe the huddling cluster of the monkeys.)

On visiting Arashiyama, Japan, on a very cold day during the winter, it is possible to observe several Japanese monkeys gathering together and making physical contact with each other in the sun. Most of these huddles include related females and their offspring, such as mothers and daughters or sisters.

On the other hand, if Tibetan monkeys were to be observed huddling and resting on a cold day, one would see various individuals, including males as well as related females, huddling together. More individuals arrive at the sleeping site in the evening and form a large huddling cluster. Even the monkeys who do not

Photo. 9-1 Huddling clusters.

usually agree with each other may make physical contact with each other and spend the night together. Even if two individuals who agree with each other make contact, a threatening male may position himself beside them. It is quite difficult for every individual in the group to join a huddle and settle down. Thus, it is natural that many events occur during the formation of the huddling clusters. How do Tibetan monkeys make such close physical contact with each other, and what kind of huddling clusters do they eventually form?

2. Observations of the formation of huddling clusters

At night, especially during the period from fall through winter and into spring, during which the weather is cold, many Tibetan monkeys made physical contact with each other and slept in a huddle on one of the rocks of the steep cliff. Everyday at around sunset, after the last provisioning, the monkeys would gather on a certain rock. Like those in other groups, the Tibetan monkeys belonging to the Yulingkeng group selected a sleeping site from among those on the cliff within their home range, depending on how far the group ranged.

Fortunately, however, when we provisioned the group during our observational periods, all the members of the group came to the same sleeping site almost everyday. I usually sat on a rock on the cliff on the opposite side of the valley and recorded the social interactions that occurred between the monkeys while they huddled and settled down for the night.

It was bitter to conduct observations on very cold days. When the weather was very cold, the monkeys finished forming their huddles before sunset. When it was relatively warm, however, they did not settle down at the sleeping site even when sunset approached. As the temperature always fell in the evenings, I shivered everyday.

The straight distance across the valley between the observer and the rock on which the monkeys slept was approximately 50 m. Even when I used binoculars with a magnification of 8, I could not identify the monkeys immediately. This task of identification became more difficult when it grew dark toward sunset. However, I continued observing them everyday and was able to identify the monkeys merely on the basis of the blurred outlines of their faces or bodies. I finally became so skilled at identifying them that I was certain of their identities. For example, I could say, "That is Huangmao, although I do not know how I am able to recognize him." I was even able to recognize the infants that were less than a year old, although I had previously been unable to identify them without marking them. However, it was still difficult for me to recognize the monkeys when I could not see their faces. When an individual in the middle of thc huddle turned his or her back to me, I would wait patiently until (s)he looked up and faced my binoculars. Whenever it turned dark, I was afraid that the monkeys would go to sleep or that I would be unable to recognize them even if they turned their faces toward me. Finally, I resorted to dropping large rocks about 20 cm in diameter from the cliff. Then, the monkeys, who were on the verge of falling asleep, would wonder what was happening and turn toward the cliff where I was seated, and I would then be able to recognize them. One snowy day, when I dropped a huge rock from the cliff, the alpha male, Huangmao, who had not joined the huddle and was wandering around, jumped out of the bushes. He jumped onto a tree growing out of the cliff and shook it violently, heedless of all the snow that was falling on him. He wanted to threaten me by asserting his power; this was one of the rare occasions on which he actually behaved like an alpha male.

The Tibetan monkeys would gather on one of the rocks of the cliff in the evening. One individual would approach another and form a huddle. Other individuals would then make contact with them and the size of the huddle would gradually increase. When individuals joined a huddle that had already been formed, they engaged in several greeting behaviors. For example, when a female approached a male, she would perform the presenting behavior. After she ensured that he had inspected her genitalia, she would make physical contact with him. Other females would perform a greeting behavior on approaching a male and then make contact with another individual beside him. A male would exhibit his penis to the dominant male or carry an infant to him for bridging, and then huddle with him. If the monkeys did not perform these behaviors, the others would stare at them, and they would be unable to join the huddle. However, an individual was rarely evicted from the huddle once (s)he huddled successfully. It was therefore a relief to the monkeys when all the members of the group had settled into the huddle. However, the alpha male often caused a stir when he attempted to mate at the sleeping site in the presence of all the monkeys. The alpha male would force a female to leave the huddle even when she was right in the center of the huddle. In such a case, the entire huddle would break up and disperse, and I would have to begin my observations again. In reality, however, this was a good opportunity to confirm the position of each monkey in the huddle, since they usually formed their huddle again within a few minutes. I had to be careful because, if I did not pay attention to this process, I would not be able to observe their positions in the huddle. After the monkeys joined the huddle, they usually groomed those who were beside them. As the huddle grew, there were some pairs who were not in actual contact with each other. Some monkeys groomed each other over another who was between them. Finally, they settled down. At this point, I completed my observations for the day, recorded the time, and descended a dark mountain path to the village of Fuxi. On one occasion, we observed the monkeys at their sleeping site until it was completely dark.

3. Which individuals in the huddle made physical contact?

Various social interactions among the Tibetan monkeys could be observed at their sleeping site, including which individuals engaged in particular behaviors and how they responded to any incidents that occurred. These interactions were so complex that 10 years had passed before I completed my analysis.

There are a few studies on primate behavior at their sleeping sites; among these are the study conducted by K. Wada on the Japanese monkeys at Shiga Height,[(1)] a study on Guinea baboons,[(2)] and a study conducted by Hiroyuki Takahashi on the Japanese monkeys at Kinkazan Island.[(3)] The wild Japanese monkeys at Kinkazan Island in Miyagi prefecture form a considerably large huddle between the rocks on a slope. I had discussed this behavior with Dr. Hiroyuki Takahashi of the Primate Research Institute, and compared it with the behavior of the Tibetan monkeys belonging to the Yulingkeng group at Huangshan.

I analyzed my data on the huddling behavior that the Tibetan monkeys

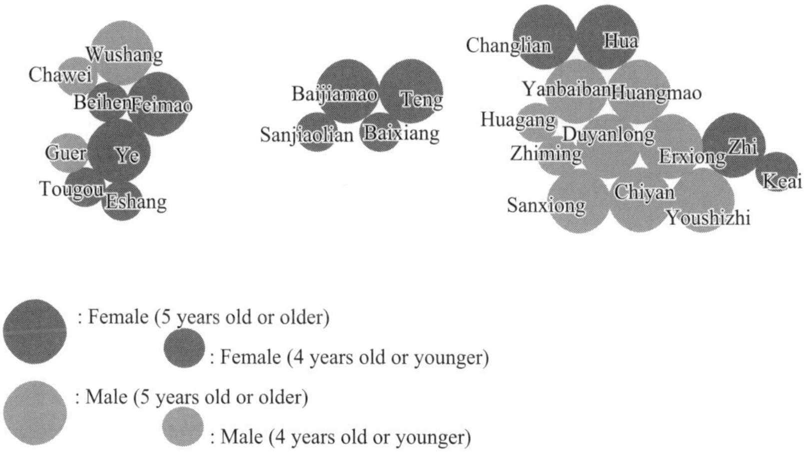

Fig. 9–1 Disposition of the huddles at the sleeping site on April 4, 1990.

displayed during the winter (from December 5, 1989 to January 21, 1990) and the spring (from March 12, 1990 to April 20, 1990). During the winter, in 19 of 42 days, and during the spring, in 28 of 37 days, I was able to observe the monkeys until they formed a huddle and finally settled down. Fig. 1–1 in Chapter 1 presents the observations recorded on January 19, 1990. Fig. 9–1 presents the disposition of the huddle as observed on April 4, 1990.

The infants were involved in several interesting events. For example, an infant sometimes jumped onto his familiar male and fell asleep beside him. The one-year-old female infants, Sanjiaolian and Keai, were usually together. However, the infants generally played around the huddles until late in the evening and finally joined the periphery of the huddle. As the infants may not exert a significant influence on the structure or composition of the huddles composed of adults, in the following section, I will analyze the behavior of seven females and eight males who were four years old or older in 1989.

I will first examine the size of each huddle. The huddles may increase in size during the winter as more monkeys try to huddle together because of the cold. In fact, the Japanese monkeys at Kinkazan Island form larger huddles as the temperature falls during the winter. While other species of monkeys usually sleep on trees, Tibetan monkeys sleep on the ground. This may be because the ground affords them more space to form a larger and warmer huddle, and also because the ground may be warmer than the trees. However, the Tibetan monkeys avoided sleeping on the ground on the gentle slopes of the mountain, which left them exposed to danger. In order to protect themselves from potential predators, they selected sleeping sites that were located on the steep cliffs.

Therefore, any given group of Tibetan monkeys will have cliffs in their home range; these serve as their sleeping sites.

On the other hand, as the weather grew warmer during the spring, the Tibetan monkeys sometimes slept on the trees growing above the cliff as well as on steep slopes. On a tree, however, even if they were to attempt to form a large huddle they would not succeed as only a few monkeys can huddle on a single branch in a line. They then resemble skewered "dangos." (Japanese "dangos" (dumplings) are usually skewered like American corn dogs.)

On snowy days, they sometimes slept in a crack between the rocks on the cliff.

I observed a total of 59 huddles during the winter and 90 huddles during the spring. On an average, the huddle was composed of 2.61 individuals (range: 1 to 7) during the winter, and 3.24 individuals (range: 1 to 10) during the spring. This difference was not statistically significant. At Huangshan, the winter is undoubtedly colder than the spring. During my observational periods in 1989, the lowest temperatures during the day in the village of Fuxi were between −2 and +7.5 degrees Centigrade, with an average temperature of 2.5 degrees Centigrade during the winter. During the spring, the lowest temperatures varied between +5 to +15 degrees Centigrade, with an average temperature of 9.3 degrees Centigrade. Although the difference between them was not significant, why was the average size of the huddles larger during the spring than during the winter? In actuality, I often forgot to record the lowest temperature of the day. If I had recorded the lowest temperature of each day or the temperature at the exact time at which the huddles were formed, my results may have indicated that the monkeys formed larger huddles when the weather was colder. While I will not attempt to excuse this oversight, I can propose another possible explanation for this interesting result. This is that the formation of the huddles reflects the differences between the social relationships of the monkeys during the mating and birth seasons. While spring is the birth season, the Tibetan monkeys continue to engage in mating during the winter. Although, by this time, the peak of the mating season that leads to reproduction has passed, many instances of mating or fights over mates occur during the winter. This may make it difficult for a large number of monkeys to huddle together.

Next, I will analyze the composition of the huddles in terms of the sex of the individuals (Fig. 9–2). There were undoubtedly more huddles that were made up of both males and females during the mating season and more huddles that were made up of females alone during the birth season. However, the difference between the composition of the huddles during the mating and birth seasons did not reveal any statistical significance. We can deduce very little about huddling behavior merely on the basis of this analysis.

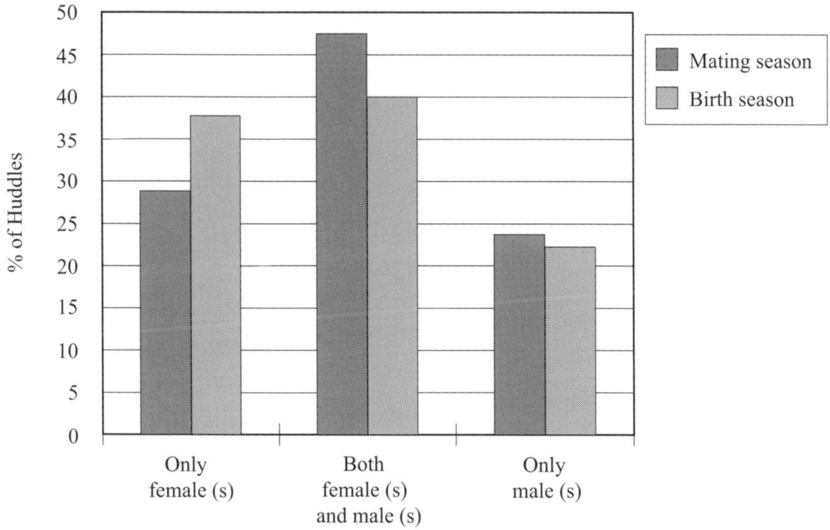

Fig. 9–2 Sex composition of the huddles.

I recorded which individuals made contact with each other as well as those who were members of the same huddle. I analyzed which individuals were beside each other in the huddle. When seen from above, a huddle appears as presented in Fig. 1–1 and Fig. 9–1, and the position of each individual is similar to that in a beehive. An individual may be in contact with 0 to 6 other individuals. An individual in the middle of the huddle is in contact with a maximum of six others. I took note of each pair of individuals who were in contact with each other and constructed a table of all these pairs, on the basis of which I compared the actual frequency and the expected frequency of contact between the individuals in each pair.

Factors such as the sex combination of the pair or the familiarity of the individuals, which may be manifested in social grooming, may have influenced which two individuals made contact in the huddle.

I will now examine the results obtained when the pairs are divided on the basis of their sex combinations. While male Tibetan monkeys huddled with other males and females huddled with females, females and males did not huddle with each other (Fig. 9–3). The frequency with which males huddled with other males and females with other females was greater than the expected frequency. In contrast, the frequency with which males and females made contact was less than the expected frequency; however, this was not significant during the mating season. The Japanese monkeys at Kinkazan Island displayed the same tendency. However, while the frequency of contact between the individuals in each male-

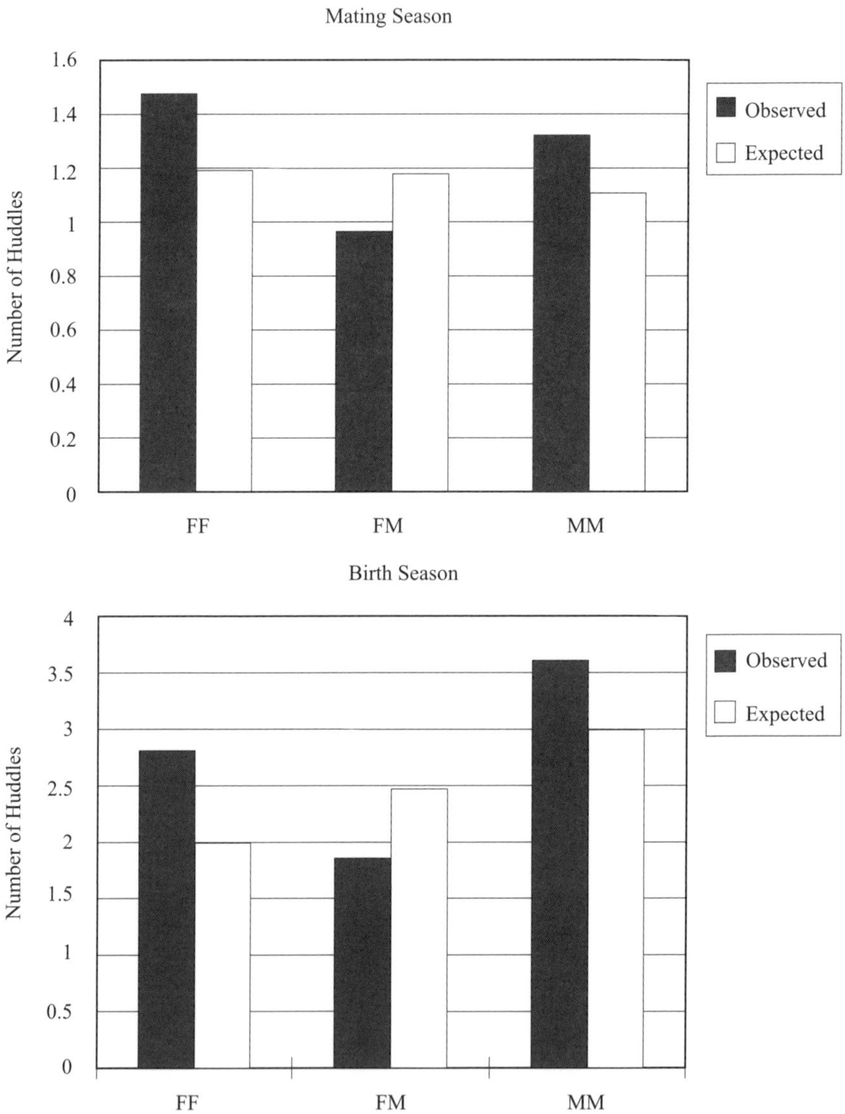

Fig. 9–3 Contact between male(s) and female(s) in huddles.
FF: Two females were in contact. FM: A female and a male were in contact.
MM: Two males were in contact.

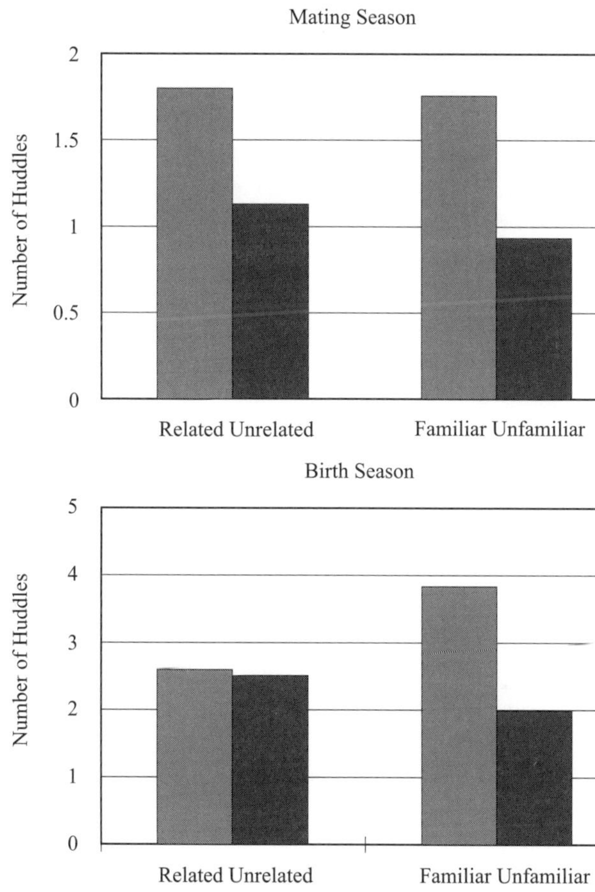

Fig. 9-4 Contact between related and unrelated pairs and between familiar and unfamiliar pairs.
Related: Mother-daughter, mother-son, and sisters (see Fig. 9-5). Unrelated: Pairs who are not related.

male pair was high among the Tibetan monkeys, it was low among the Japanese monkeys. The familiarity between male Tibetan monkeys is more evident than that among male Japanese monkeys.

Further, our results did not indicate that the frequency with which related Tibetan monkeys (mothers and daughters or mothers and sons, and maternal siblings) huddled together was significantly greater than that among unrelated monkeys (Fig. 9−4). Unlike Japanese monkeys, who were more likely to huddle with individuals who were related to them, this is a characteristic feature of Tibetan monkeys. These differences between Japanese and Tibetan monkeys correspond to the differences in their social interactions during the day.

Do two individuals who engage in social grooming during the day make contact with each other in the huddling clusters as well? The answer is "Yes"; those who groomed each other with more than the mean frequency huddled more frequently than those who groomed each other less often (Table 9–4).

I would like to direct the reader's attention to Fig. 9–5, in which I have used straight lines to connect the individuals who huddled more frequently than was expected at the sleeping site. Fig. 9–5 clearly reveals the characteristics of the social relationships among the Tibetan monkeys during the study period. For example, Duyanlong and Sanxiong huddled frequently during the winter; this is indicated by the thick lines connecting them. During the subsequent spring, the males who had successive ranks, Duyanlong and Sanxiong, the low-ranking new immigrant males, Yanbaiban and Chiyan, and Erxiong and Youshizhi, who immediately preceded them in rank, also huddled frequently. On the other hand, high-ranking and low-ranking males, such as Huangmao and Yanbaiban, Huangmao and Chiyan, and Wushang and Duyanlong, huddled in the winter, and Huangmao and Yanbaiban, Wushang and Duyanlong, and Wushang and Sanxiong, huddled in the spring. The alpha male, Huangmao, and the second-ranking male, Wushang, did not huddle with each other. In contrast, it is interesting that the alpha female, Baijamao, and the second-ranking female, Zhi, huddled together. This may indicate that during this period, the highest-ranking male had a hostile relationship with the second-ranking male, while the alpha female was allied with the second-ranking female. With regard to contact between the males and females, the alpha male Huangmao and certain females huddled together. However, the males and females who were in a long-term consortship necessarily huddled together at the sleeping site.

4. Which three monkeys huddle together?

The following is the most important section of this chapter. Here, I will examine which combinations of three individuals made physical contact with each other and formed a triangular huddle at the sleeping site. This is the most curious aspect of huddling behavior, because I have studied social behavior of monkeys in the light of the interactions among three individuals—that is, bridging, the formation of coalitions, and the interactions between twin Japanese monkeys and their mother. It is natural to assume that if A and B, B and C, and C and A frequently associate with each other, the three individuals A, B, and C will assemble and form a triangular huddle. This implies that the sets of three individuals connected with thick lines in Fig. 9–5 should frequently form

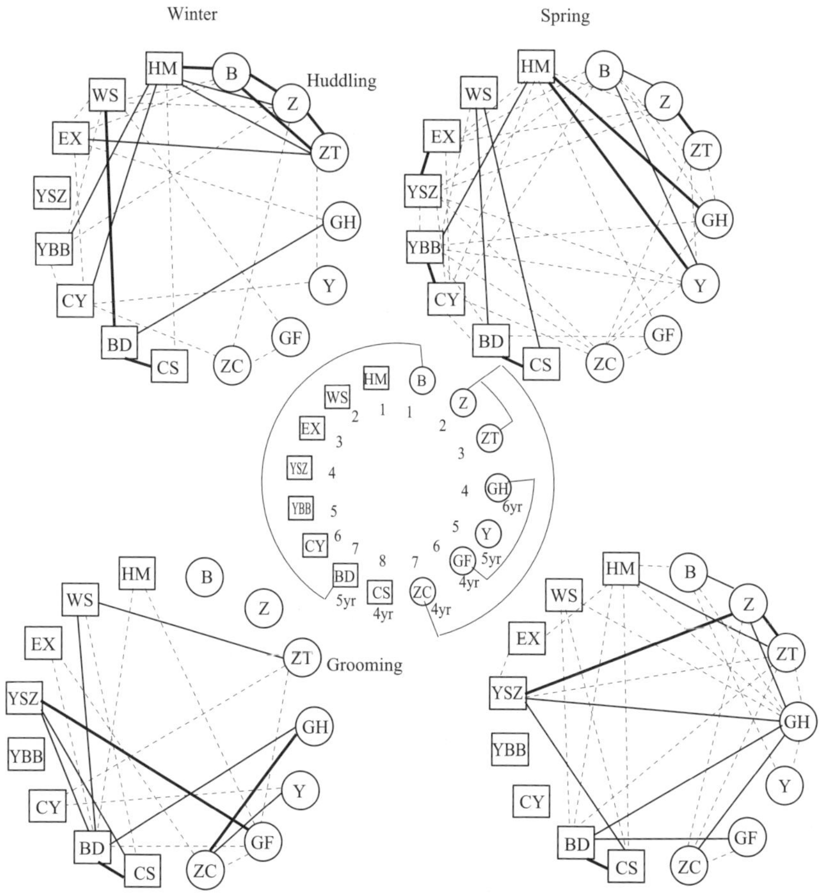

Fig. 9–5 Frequency of the formation of dyadic huddles and social grooming.
The lines indicate the relative frequency of the formation of dyadic huddles at the sleeping site and social grooming during the day. Two individuals are connected by lines if the frequency of huddling and social grooming between them was greater than the mean X (X $\leq$ dotted line < SD, SD $\leq$ thin line < 2SD, and 2SD $\leq$ thick line). The names of the individuals are enclosed in circles (females) and squares (males). Females are arranged in the descending order of their dominance ranks, clockwise from B, and males are arranged in the descending order of their dominance ranks, counterclockwise from HM. In the central diagram, kin-related individuals are connected by arcs and the ages of monkeys are noted if they were less than 8 years old in 1990.
Names of the monkeys:
B=Baijiamao. Z=Zhi. ZT=Teng. GH=Hua. Y=Ye. GF=Feimao.
ZC=Changlian. HM=Huangmao. WS=Wushang. EX=Erxiong
YSZ=Youshizhi. YBB=Yanbaiban CY=Chiyan. BD=Duyanlong.
CS=Sanxiong.

triangular huddles. However, it is possible that additional factors play a role in determining which three individuals make contact with one another.

For example, in the case of interactions among three people, there may exist certain situations in which any two of the three are on good terms with each other, but all three cannot concur. On the other hand, the individuals in a group of three people that does not include a pair of good friends can get along when they are together. The group of three individuals may share a certain compatibility that cannot exist in a dyadic relationship.

Let us consider a social situation involving three or more individuals to be a system wherein the social interactions between the individuals mutually and simultaneously determine each individual's behavior. If every characteristic and the initial condition of each individual are known, it may be possible to predict the interactions between two individuals. This is similar to the way in which it is possible to predict the future motion of two objects on the basis of gravity and magnetic force by understanding their characteristics, initial position and the direction and speed the movement. However, it is difficult to predict the future motion of three objects. This is because each object mutually and simultaneously influences the other objects, and the three give rise to a complex system. Thus, the interactions among three individuals cannot be predicted like those between two individuals. It is still unknown whether three individual nonhuman primates can share a compatibility that cannot exist in a dyadic relationship. However, this may be ascertainable by studying the way in which each individual's decision-making process exerts a mutual and simultaneous influence on those of the others, and how the compatibility emerges from the interactions among the three individuals. This issue of the compatibility among three individuals can be examined by exploring the interactions among three or more individuals based on the social interactions observed among primates. It is my ambition to study this topic.

As the first step toward achieving this goal, I attempted to find three individuals who huddled together with more than the expected frequency. There may be three monkeys who are more likely to form a triangular huddle than the expected frequency calculated from the frequency of the contact between two individuals in the disposition of the Tibetan monkeys at the sleeping site. Alternatively, there may be three monkeys who form a triangular huddle less frequently than expected. In addition to triangular huddles that are composed of only three individuals, several triangular huddles may also be formed within a larger huddle. If a triangle is formed within a larger huddle, another individual usually comes into contact with this huddle, resulting in the formation of multiple triangular huddles within the large huddle. Is there a possibility that while counting the triangular huddles, three individuals who accidentally happen to be in a triangle may also be included? Further, an individual who arrived at the site late would be unable to make contact with the individual at the center of the huddle without forcing his or her way through or climbing over another's head. Nevertheless, the new arrivals at the sleeping site gradually joined the huddle that had already been formed. Those who approached the huddle usually contact two individuals who were already a part of it. That is, the Tibetan monkeys chose two individuals to make contact with, and therefore, any triangle was the result of the choices of those who were the last to join the huddle.

First, I counted the number of triangular huddles formed by each combination of three individuals.

Second, I estimated this likelihood on the basis of the frequency with which three dyads in the triad were formed respectively. When three individuals are chosen from among the individuals A, B, C, D, E, F..., the probability that those three will be A, B, and C is {(the probability that A and B are chosen) × (the conditional probability in which C is chosen when A and B are chosen)}. Here, I assume that the conditional probability in which C is chosen when A and B are chosen is determined solely by the likelihood of contact between C and A, and C and B. This means that it is not necessary that C is especially likely to be involved only when A and B are chosen together. I assume that the likelihood of C being involved is the arithmetic mean of the likelihood with which A is chosen but B is not, and with which A is not chosen, but B is. That is, the function presenting the probability that C is chosen is the linear function as follows:

$p \times F(x) + q \times G(y)$,
p and q: fixed numbers
F(x) and G(y): functions
x and y: the output of the relationship between C and A, and the output of the relationship between C and B, respectively.

Remark: When three elements are chosen from a certain group composed of A, B, C, D, and so on, [the probability that A, B, and C are chosen simultaneously] is [the probability that A and B are chosen simultaneously] multiplied by [a conditional probability that C is chosen when A and B are chosen simultaneously.]. [The probability that C is chosen] may increase or decrease when both A and B are chosen simultaneously, as compared with when either A or B is chosen separately. Here, however, [the conditional probability that C is chosen when A and B are chosen simultaneously] is assumed to be the arithmetic mean of [the conditional probability that C is chosen when A is chosen and B is not] and [the conditional probability that C is chosen when B is chosen and A is not]. That is,

$p(ABC) = p(AB) \times p_{ab}(C) = p(AB) \times 1/2\{p_{a\bar{b}}(C)/2 + p_{\bar{a}b}(C)/2\}$.

p(ABC): the probability that A, B, and C are chosen simultaneously.

p(AB): the probability that A and B are chosen simultaneously.

$p_{a\bar{b}}$ (C): the conditional probability that C is chosen when A is chosen and B is not.

Finally. I compared the expected value obtained by this assumption with the actual values, and found that there was a statistically significant difference between the observed value of the combination of the three individuals that formed a triangular huddle and made contact with one another, and the expected value obtained from the frequency with which two individuals huddled. This indicates that other factors besides the frequency with which two individuals were likely to huddle exert an influence on which three individuals huddle together. This means that individual C did not determine whether or not to contact both of A and B on the basis of a linear function regarding the relationship between him/herself and A, and the relationship between him/herself and B.

Unfortunately, however, I could not concretely show which three individuals huddled together. Seven females and eight males yield a very large number of combinations, that is, 455 combinations of three individuals. The actual frequency of each contact is 0 in most triads. For example, while the actual frequency of the formation of a triangular huddle composed of Baijamao, Zhi, and Teng was actually 2, the expected frequency was 0.29. Since I had no other idea, I classified each individual on the basis of categories such as sex, and analyzed which categories would display frequencies that were higher or lower than the expected frequency.

If [+] denotes a pair of individuals who frequently huddle together and [-] denotes a pair who are less likely to huddle, then there will be four possible combinations: [+++], [++-], [+- -], and [- - -]. Considering the likelihood with which each dyad in the triad huddled, which three individuals did frequently huddle together? The results revealed that the individuals belonging to a triad in which all the dyads huddled more frequently were more likely to huddle together. It is a matter of course for the three individuals belonging to a triad in which all the three pairs are likely to huddle frequently to huddle together. However, the likelihood of a triad including three pairs who frequently huddle together was accelerated beyond the prediction. The triad including three pairs who frequently huddle together huddled more frequently than expected (Fig. 9–6). When the triad included two individuals who were less likely to huddle together, the three individuals in the triad were also less likely to huddle together

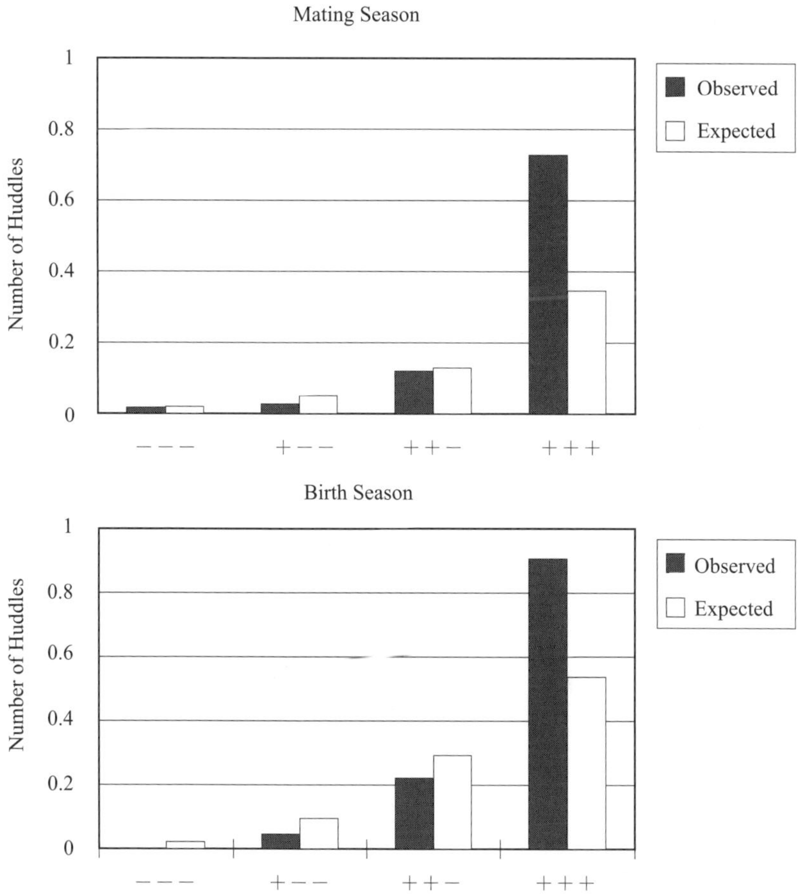

Fig. 9–6 Frequency of the formation of triadic huddles for different dyadic affiliations. The figure presents the means of the observed and expected (see Chapter 9) number of triangular huddles formed at the sleeping site during the winter and spring study periods. The affiliations of dyads within the triad are indicated by [− − −], [+ − −], [+ + −], and [+ + +]. For instance, [+] implies that the dyad formed dyadic huddles with more than the mean frequency. [+ + −] implies that the triad was composed of 2 dyads that formed dyadic huddles frequently and 1 dyad that did not.

than expected. The same result held true for Japanese monkeys.

I will now examine the combinations according to the sex of the individuals. The combinations of three individuals based on their sex can be classified as [FFF], [FFM], [FMM], and [MMM] (M: male; F: female). Is it a fact that an odd number of women is not working well? Fig. 9–7 presents the average number of times each triad huddled according to the different combinations of the three

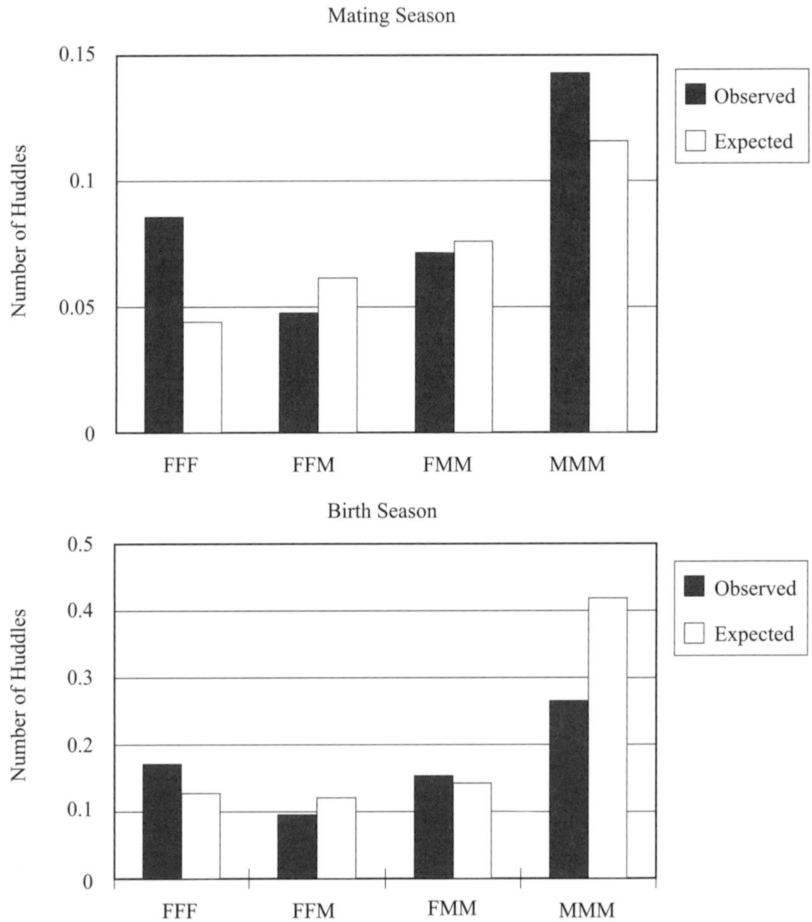

Fig. 9–7 Frequency of the formation of triadic huddles of females and males. The figure presents the means of the observed and expected (see Chapter 9) number of triangular huddles formed at the sleeping site during the winter and spring study periods. Triangular huddles consisted of the following combinations: FFF (female-female-female), FFM (female-female-male), FMM (female-male-male), and MMM (male-male-male).

individuals. These categories are [MMM], [FFF], [FMM], and [FFM], in decreasing order of the average number of times they huddled.

However, these results merely reflect the fact that males and females are less likely to huddle together. They do not indicate that, for example, the combination [FMM] is less likely to be formed. The results obtained by comparing each observed value to the expected value indicate that during the mating season, the frequency with which Tibetan monkeys huddled in the combinations [FFF] and [MMM] was greater than the expected frequency, and the frequency with which they huddled in the combinations [FFM] and [FMM] was less than the expected frequency. During the birth season, the combinations [FFF] and [FMM] were more likely to be formed, and the combinations [MMM] and [FFM], less likely, though there was no significant difference in the case of the combination [FFF]. Among Japanese monkeys, the combinations [FFF], [FFM], and [FMM] were less likely to be formed.

Three female Tibetan monkeys were more likely to huddle than expected, while three female Japanese monkeys were less likely to huddle than expected, though this was not significant. Among Japanese monkeys, the tendency of unrelated females to avoid huddling may have reduced the likelihood of females huddling together, while the results obtained in the case of Tibetan monkeys may reflect the absence of this factor.

The results of my analysis revealed another tendency. I observed frequent formations of one-female and two-male triads during the birth season. (I named this type of formation the "Dreams Come True formation" after a Japanese music band called "Dreams Come True," which comprises one female vocalist and two male players.) During the birth season, the number of formations of female-male-male groups of Tibetan monkeys was greater than the expected value. During the mating season, the frequency of this type of formation was not more than the expected value.

I analyzed the frequency with which males or females are the last individuals who participated in triangular huddles. As a result, for example, the number of formations of the [FFM] group is small because the females did not approach male-female dyads, while males frequently approached female-female dyads. The combination [FMM] was not often formed because males usually did not approach male-female dyads. Females approached male-male dyads quite often.

What are the reasons behind these tendencies?

The [FMM] triad might not be formed frequently because of the competition between the males for the females. Tibetan monkeys continued mating during the winter observational period. As a result, the competition between the males for the females was more intense during the winter than during the birth season in the spring. It is interesting that this competition was reflected neither by the males not huddling with other males nor by three males not huddling together. It was, however, reflected by the fact that one female and two males did not huddle together. Two males or three males would huddle together, but males would not approach male-female dyads. During the birth season, this negative factor disappeared and the monkeys formed huddles composed of one female and two males.

Computer simulations have been used to study the behavior of gregarious birds and fish, that is, how each individual determines its movement within the group and how this generates the movement of the group as a whole. In these studies, the movement of a group can be reconstructed on a computer if each individual's behavioral standard is appropriately set in terms of the position and movement of another individual. Can the formation of huddles among primates be similarly reconstructed if we set appropriate standards for their behavior? Since primates determine their behavior not only on the basis of the relationship between themselves and others but also on the basis of the relationships between other individuals, we cannot understand every aspect of their social interactions by merely observing the relationship between two individuals. We should assume that primates determine their own positions in a huddle according to their knowledge of the characteristics of other individuals and the relationships among them. If we determine the nature of the analyses and the technology of the simulation on the basis of the results obtained from the comparison of the simulations and the actual research, we may be able to understand the way in which a certain kind of compatibility among three individuals emerges from their social interactions. I expect that this could provide us with a clue to solve the problems in human relationships, including that of the compatibility among multiple people, which has not yet been scientifically understood. Moreover, I expect that the results and methodology obtained from such research can help us understand, for example, the way in which a particular type of subgrouping is created by the fission-fusion social groups among chimpanzees, the way in which the direction of the movement of groups of primates is determined by the individuals in that group, and the way in which to determine the disposition and composition of grouping so as to generate smooth human relationships at the workplace or other public places.

Chapter 10

How Do Primates Understand the Mind and Behavior of Other Individuals?

1. Deception on the part of primates

Observational case: Beihen ran around

One of the unforgettable events I observed involved Beihen. It occurred on a beautiful snowy day in 1989. Beihen, Wushang, and Duyanlong were at the provisioning site. While Duyanlong was still a juvenile, low-ranking male, his rank was higher than that of three-year-old Beihen. Therefore, these individuals, in the ascending order of their ranks, were Wushang, Duyanlong, and Beihen. They were looking for corn in the cracks between the rocks, which were covered with snow. I wished that Beihen would do something interesting while I was observing her. My wish may have been granted because Beihen suddenly began screaming at Duyanlong and running around him. Wushang appeared to be bewildered by this and stared at Duyanlong, whose rank was much lower than Wushang's. Although Duyanlong was not at fault, he was, unfortunately, unable to communicate this to Wushang. Wushang chased Duyanlong, who disappeared into the woods. Beihen remained at the provisioning site as though nothing had happened, and began foraging for corn in the areas that had been occupied by Duyanlong and Wushang. Wushang returned a few minutes later. Duyanlong returned much later, as though nothing had happened; however, he was completely covered with snow.

At that time, Wushang had formed a consortship with Beihen. Since she was still a juvenile, she often jumped onto small branches and avoided Wushang when he attempted to mate with her. However, in this instance, Beihen used Wushang's support to defeat Duyanlong.

When Beihen used Duyanlong's support to secure a higher rank in 1990, she often cried even when nothing had happened to her. Therefore, any male who heard her would come to her rescue. During this period, the male who came to her rescue was Duyanlong, and he was manipulated by her. If monkeys determine their future behavior on the basis of the assumption that "If I do this, this individual will do that," they can manipulate the other individual's behavior. There are several behaviors that can be used to manipulate others, and even a weak individual can indirectly manipulate the behavior of others by altering their mental states by giving them certain information or deceiving them. Beihen was adept at deceiving and manipulating the males.

Cases of deception have been reported among group-living primates.(1) For example, on one occasion, an infant baboon (*Papio ursinus*) suddenly screamed at a juvenile male who was pulling at and eating bulbs beside him although he had not bullied or provoked the infant. The infant's mother then arrived and chased the male away, and the infant was able to eat the bulbs that the male had already pulled out of the ground.

Another incident involved a male baboon (*Papio ursinus*), who was chased away by angry females because one of their infants had screamed at him. While he was being chased, he suddenly stopped running, stood on his hind limbs, and stared in a certain direction as though trying to spot a dangerous predator. The females stopped and looked in the same direction, and thus stopped chasing him.

Another incident involving Hamadryas baboons (*Papio hamadryas*) was observed. An adult female sat beside a juvenile male who did not belong to the group. A resident male of the group would intervene if he noticed one of his females grooming a male who was an outsider. However, this female moved to and sat behind a rock, where a resident male would be able to see her head and back but not the outsider and the front of the female's body and her arms. She then began grooming the outsider.(2)

Another incident involved chimpanzees in an enclosure. Two adult males, A and B, were fighting over a higher rank and attempted to prove their power. A, who was chased by B, climbed onto a tree, and turned his back toward B. When B screamed and challenged him again, A bared his teeth and made an expression resembling a grimace. This is the expression that subordinate males make in front of the dominant male when they have been attacked and have lost the fight. At that moment, A pressed his fingers to his upper and lower lips. However, his facial expression was still a nervous grimace. It was not until he succeeded in effacing his nervous expression with his fingers that A turned to face B.(3)

E. W. Menzel(4) and T. Matsuzawa(5) constructed a social situation in which deception was possible and useful to the chimpanzees in an enclosure. First, they showed chimpanzee A the place in a field where they had hidden bananas. They then returned individual A to the cage with the other chimpa-nzees, and then released all the chimpanzees into the field. After repeating this procedure several times, it was observed that each individual's behavior kept changing. Initially, individual A, who knew the location of the bananas, went directly there to get them. Next, the other chimpanzees followed A, who was going toward the bananas, and thus obtained those bananas. Third, A did not run to the place where the bananas were hidden, but made a detour or sat down and waited. He thus prevented the other chimpanzees from stealing the bananas and secured them for himself as soon as the others ran in different directions.

In cases such as the ones above, in which each individual's interests are in conflict, the individual who sends out certain information can secure his or her own advantage by producing signals that will manipulate the receiver of this information. (S)he can even deceive the receiver by sending out false signals that may mislead him or her. However, the receiver may respond to the sender of the information so as to avoid being deceived. For example, assume that in the case of competition over a resource, an individual sends a signal that implies, "I will attack you," although (s)he does not actually intend to fight over the resource. If the sender uses this signal to deceive the receiver, (s)he would be able to secure the resource effortlessly. However, the receiver may not believe this information since the sender can produce false signals expressing an intention to attack the receiver without incurring any cost.[(6)(7)] A male's attempts to attract a female and ensure that he is chosen for mating by pretending that he has better genes than he actually does will entail a cost to the male. This is because the weaker the male is, the more difficult it is for him to attract a female by trying to make his appearance more appealing than it actually is. As a result, the male is compelled to send out truthful signals and the females can evaluate him on the basis on that signal.[(8)] Thus, deception is difficult. When the receiver cannot distinguish between true and deceptive signals, however, (s)he can be deceived by the sender. Many species of birds, such as chickadees, make alarm calls announcing the presence of hawks. However, they occasionally make false alarm calls in the absence of predators. This is done to chase the others away so that they can monopolize the food.[(9)] However, this type of deception is not always effective. When individuals send out false signals, they also incur a risk. In the case of provisioned rhesus monkeys, the individuals who find food sometimes inform the others by making certain calls, while at other times, they do not inform the others. However, if the other monkeys find that a certain individual has concealed the location of food, that individual is often attacked.[(10)] Further, an individual who frequently sends out false signals will eventually be ignored by the receiver. In a certain experiment, researchers recorded a wild Vervet monkey's alarm calls for hawks, and using a tape recorder, repeatedly played these to the other members of the group when there was no hawk in the vicinity. The monkeys that had heard these repeated alarm calls stopped paying attention to them. They still responded to the same individual's alarm calls for other predators and paid attention to other individuals' alarm calls for hawks.(11) That is, they stopped believing a particular type of alarm call produced by a certain individual. I sympathize with the Vervet monkey who became "the boy who cried wolf" because of this experiment. The receiver should ignore such signals, even though it is possible that they could be true sometimes, if the proportion of true signals to false signals exceeds the benefits obtained by responding to honest signals or the losses sustained by responding to false signals. The sender can expect false signals to be effective only when (s)he rarely produces them. This holds true as long as the receiver does not incur heavy losses by responding to false signals, and obtains significant benefits by responding to true signals. Ultimately, the receiver decides whether or not to respond to a signal by determining the costs and benefits involved and also on the basis of his or her ability to distinguish between true and false signals.(12)

If the receiver has exceptional social abilities, (s)he will be able to judge whether the signal is false on the basis of the social context, even when (s)he cannot differentiate between true and false signals on the basis of their physical characteristics. Once the receiver tries to detect deception on the basis of such a clue, the sender, too, makes false signals according to the social context instead of producing fewer false signals. As seen in Fushimi's experiment on primates (in Chapter 2, Section 3), the sender of the information can predict the receiver's response and select signals that will make the receiver behave in a manner that is advantageous to the sender. The sender can successfully deceive the receiver by manipulating his or her predicted response.

Humans deceive others by judging whether or not the other person possesses certain knowledge. In order to successfully deceive another individual, we think, "This person does not know about this, so he must believe me if I tell him this." Do other animals, like humans, actually deceive others by making assumptions regarding the other individual's knowledge?

2. Theory of mind

People often wonder what animals such as cats and dogs think and feel. Let us assume the case of the dog named "Koro." Suppose that when you open a cabinet to look for dog food with which to feed Koro, you cannot find the food because one of your family members has moved it. Koro first whimpers, but later begins to bark. You may think "Koro knows where the dog food was put after his breakfast." "Is Koro indicating that the dog food is not in the cabinet?" You may wonder whether Koro is thinking, "Why are you not giving me my dinner? Don't you know that there is no dog food in the cabinet you are looking in?" In this situation, is Koro actually wondering what you are thinking, just as you are wondering about the mind of Koro?

Do monkeys or dogs really wonder what humans or other individuals of their own species think? Koro may bark merely because he knows that "If I bark in front of this large animal, I may be given my meal." When animals display behavior that resembles that of humans, we cannot assert that such behavior occurs because their minds have the same mechanisms as those of humans if we can explain their behavior by simpler mechanisms.

Do animals understand the minds of other individuals? This research topic has its roots in D. Premack and G. Woodruff's article in 1978, titled "Does the chimpanzee have a theory of mind?"(13) In this article, they argued that "An individual has a theory of mind if he imputes mental states to himself and others. … for example, purpose or intention, as well as knowledge, belief, thinking, doubt, guessing, pretending, liking, and so forth." Having a theory of mind implies that the individual can understand the minds of others. The authors used the term "theory" because just as the movement of an object can be predicted using a theory of physics, animal behavior, too, can be predicted using a theory of mind, although a mind cannot actually be observed.

Let us delineate the way in which humans use a theory of mind to predict the behavior of others. When humans predict the behavior of others they, (1) recognize that the other individual is not an inanimate object but has the ability to move voluntarily and that they should not apply a theory of physics to the behavior of animals, and (2) utilize the principle that the decision-making mechanism of others is basically the same as their own. They believe that when another individual is faced with a situation identical to their own, in general, it is reasonable to believe that the other person will behave the same way as they do. To predict the behavior of others, humans use analogy that "I behave in this way under this particular circumstance. The other person, too, behaves in the same way." Further, they usually think, "Under this circumstance, I behave in this way because I feel this way. Since the thoughts and feelings of the other person under this circumstance are the same as my own, the person will behave in this way."

However, the circumstances that other people are faced with are usually different from our own. Things that an individual sees depend on the location of the other individual. In this case, a theory of perception that explains "how things are seen by other individuals" is required. Further, it is necessary to understand that the knowledge that others possess depends on their perception and to estimate their knowledge accordingly. This is what a theory of knowledge entails. Therefore, in (3–a), we can estimate the other individual's knowledge by understanding his or her perception, which may differ from our own.

Further, the character of the other person may also differ from our own, and (s)he may experience different feelings or desires under certain circumstances, depending on his or her character. In this case, we need a theory of feelings that explains "what and how much the person wants." Accordingly, in (3−b), we can estimate the other person's desires and intentions.

By combining (3−a) and (3−b), the other person's behavior can be predicted in (3−c), that is, (s)he sees and knows a certain thing, wants to do it, and therefore, (s)he does it. In order to apply a theory of mind to other individuals and predict their behavior, it is necessary to understand their state of mind at each stage and the relationships between those stages.(14)

While examining the ways in which animals such as chimpanzees, which are closely related to humans, predict the behavior of others, it is possible to use a methodology that experimentally selects each step of the process of behavioral assumption in humans.

H. Winner and J. Perner formulated an effective method of conducting experiments on the ability to understand the knowledge of other individuals, and evaluated this ability in human infants.(15) This is known as a false belief task. In an experiment involving a false belief task, the following story was presented to the subjects (human infants) by means of a puppet show or a videotape: A boy named Maxi helped his mother to make a chocolate cake and they put the chocolate on a green shelf. Later, Maxi went outside to play. While he was outside, his mother removed the chocolate from the shelf and used some of it. She then put it on a blue shelf instead of the green shelf and left the kitchen. Maxi returned and wanted to eat the chocolate. The subjects were then asked the following question: Where does Maxi look for the chocolate? If the subjects were able to understand the difference between the knowledge possessed by the character, Maxi, and their own knowledge, they would be able to answer, "Maxi looks for the chocolate on the green shelf." Normal human infants are able to arrive at the correct answer in the false belief task when they are approximately four years old. Further, when they are eight or nine years old, they are able to understand a secondary belief, namely, "Mother thinks that Maxi believes the chocolate is on the green shelf."

D. Premack et al. constructed a similar situation for the chimpanzee, Sarah.(16) She was presented with two boxes, one containing a "good" object, a banana, and the other a "bad" object, a rubber snake. Everyday, Sarah's caretakers would open one of the boxes; they usually opened the one containing the good object. One day, a stranger entered and switched the contents of the two boxes while Sarah was watching, but the caretaker was absent. What did Sarah do the next time the caretaker entered? The structure of this situation is the same as that of the false belief task in which the other person's knowledge is different from one's own. Sarah knew that the contents of the boxes had been switched; she was aware of this as she had watched this process with anger. However, when the caretaker entered, Sarah's behavior was no different form the usual. In this experiment, the chimpanzee did not display an understanding that the knowledge possessed by other individuals was different from her own.

However, the chimpanzee, Sarah, succeeded in the following experiment.(13) She was first shown a video of a person trying to obtain a banana hung from the ceiling, which was out of the person's reach. Later, Sarah was asked to select one of two or three pictures presenting solutions to the problem encountered by the person in the video. In this experiment, she selected a picture depicting the person standing on a footstool. It is likely that Sarah was able to understand the problem encountered by the person and what the person intended to do.

Further, D. J. Povinelli et al. conducted the following experiment.(17) Chimpanzees were shown some food being put in a box. However, as there were blinds between the box and them, they were unable to see in which out of several boxes the food was being put. Two people stood behind the blinds; one person could see which box the food was put in while the other was blindfolded and unable to do so. The chimpanzees were made to open any one of the boxes according to the directions of these two people. Three out of four chimpanzees opened the box indicated by the person who had been able to see which box the food was placed in. This suggests that these chimpanzees understood the relationship between "seeing" and "knowing." They understood that the person who was able to see was the one who possessed the knowledge.

The results of these experiments conducted on chimpanzees suggest that they possess a theory of feeling that is similar to that of humans. They understand the intentions of other individuals and have a theory of perception that allows them to understand what others see. However, with regard to a theory of knowledge, while they were able to understand that perception affects knowledge, they were unable to understand that the knowledge possessed by other individuals is different from their own.

Nevertheless, in Premack's and Matsuzawa's experiments, the chimpanzees behaved as though they possessed a theory of knowledge in the interactions between those who knew where the bananas were hidden and those who did not. Satoru Hirata is currently examining this issue using a similar experiment. A theory of knowledge covers four possible cases, that is, a case in which both other individuals and the individual in question possess certain knowledge, a case in which other individuals know something that the individual in question does not know, a case in which other individuals do not know something that the individual in question knows, and a case in which neither of them know. It is interesting to examine the individual's understanding of each situation. In addition, it is remarkable if chimpanzees are able to understand a secondary belief such as "I know that you know my thoughts."

D. C. Dennett suggested that the behavior of individuals could be categorized into several levels according to the individual's intentionality.[(18)] He defined level 0 as the case in which the individual's behavior is unintentional. Level 1 is the case in which the individual's behavior is intentional. That is, the individual expects that the behavior will produce certain results. Level 2 is the case in which the individual is aware of another individual's intentions. This can be possible only when the individual has a theory of mind.

On the basis of this classification, R. Byrne and A. Whiten examined the various kinds of "deception" practiced by animals.[(1)(23)]

One of the best known deceptions among animals is mimicry, such as the imitation of the shapes of the wings and color patterns of insects.[(19)] This type of deception is categorized as level 0 because it involves species-specific forms and behavioral patterns, which are genetically programmed to result in the deception of other individuals.

Among European quails (*Perdix perdix*), the mothers protect their babies from predators by pretending that their wings are broken, thereby drawing the predators' attention toward themselves and away from their babies. In this case, it is difficult to assert that the mother birds have certain goals in mind. However, I believe that, for example, Beihen had a certain goal in mind, namely, to induce Duyanlong to attack her enemy, and that she understood that in order to achieve this goal, she would have to scream in a certain situation. The infant baboon belonging may also have had a goal, namely, to induce his mother to drive the male away, and may have expected his behavior to produce a particular result. This is suggested by the fact that all baboons do not scream in these circumstances, and the infant screamed only when his mother was not watching him.

Further, can the infant's behavior in this example be interpreted as the result of his expectation that if he screamed, his mother would believe that he was being bullied? However, the infant may have merely known that when somebody subordinate to his mother monopolized the food, he himself would be able to secure the food if he screamed when his mother was close enough to hear his voice but was not looking at him. On the basis of this knowledge alone, he may have performed this behavior without making any assumptions regarding his mother's mind. Therefore, we cannot assert that this behavior is level 2 merely on the basis of observational facts. In fact, it has been reported that rhesus monkeys could not understand the relationship between seeing and knowing. This was discovered in an experiment similar to the one previously discussed, in which chimpanzees — in this case, the rhesus monkeys — opened a box containing bananas.[21] When the female hamadryas baboon groomed in a spot where she could not be seen by the resident male, she seemed to know that even though she was near the resident male, she could groom other males since the resident male could not see her. This case may indicate an understanding of the perceptions of other individuals, that is, the way in which others see things. This step may be a precursor to the understanding of the minds of others. Nevertheless, monkeys do not appear to be able to use the gaze of other monkeys.[22] If an individual were to indicate that a box contains a banana by gazing in the direction of the box, humans would be able to understand that the box contains a banana on the basis of the direction of the gaze. However, the monkeys were not able to use the gaze of other monkeys. This is one of because gazing into an individual's eyes indicates a threat in the case of monkeys. Because nonhuman primates and humans have different characteristics and mental process, we may not be able to evaluate the ability of nonhuman primates by experimentally examining if nonhuman primates have the same ability that human use in each step of a theory of mind.

R. Byrne and A. Whiten accumulated various cases of deception among animals from researchers who conducted observations on animals.[23] Taking the observational time into consideration, they examined which kind of deception was observed in each animal species. Their research showed that level 1 deception was not observed among primitive primates, that is, prosimians. Level 1 deception was observed among anthropoids, including monkeys and apes. Deception that involves an understanding of the perceptions of others was observed among old world monkeys, that is, cercopithecidae, and anthropoids such as great apes. It appeared reasonable to interpret certain observational cases as examples of level 2 deception that is intentional and involves the assumption of the state of mind of other individuals. All the cases of such deception, with the exception of one, were observed among great apes.

We can use "teaching" in the same way as "deception" for the study of mind reading of animals. From a functional viewpoint, the behavior of the individual A is regarded as teaching if A modifies his or her behavior only in the presence of individual B and this behavior causes B to obtain more efficient knowledge or skills although this behavior does not benefit A. If this behavior is defined as teaching, then teaching can be observed among raptores, carnivores, and primates.[23] Among cats, mothers allow their infants to practice their hunting skills by giving them prey such as a mouse. Among carnivores and raptores, parents encourage youngsters to develop their hunting skills in a similar manner. Among primates, teaching may assume the form of providing several opportunities to learn, encouraging appropriate behaviors, or suppressing inappropriate behaviors. For example, among rhesus monkeys and chimpanzees, mothers encourage their infants to walk by gradually retreating when the infants approach them. Among Japanese monkeys, mothers pull their infants away from dangerous objects, and among chimpanzees, mothers try to remove any food that is not normally eaten from their infants' mouths.

Wild Vervet monkeys make alarm calls for predators more frequently when they are with others than when they are alone. This means that they make alarm calls because they have someone whom they intend to teach.

However, in order to judge whether or not the teacher understands the state of mind of other individuals, it is necessary to show that (s)he sends certain signals because the other individuals are not aware of the situation. It was examined whether the mothers among provisioned Japanese monkeys and rhesus monkeys would change the frequency of their alarm calls depending on whether or not their infants could recognize dangerous objects.[25] In this experiment, the Japanese and rhesus monkeys did not change their behaviors in order to teach other individuals something according to whether or not the other individuals possessed certain knowledge.

Thus, on the basis of observations and experiments on interactions involving "teaching" or "deception," we could not obtain proof that primates other than great apes and humans change their behaviors depending on the knowledge of other individuals. In this book, we have seen that primates understand that other individuals' social relationships which are different from their own, and have the ability to select their own behaviors based on predictions of others' behavior and their ability to manipulate others' behavior. However, although primates are able to understand that the social relationships of others are different from their own, they do not understand that the minds of others are different from their own. When they manipulate or predict the behavior of others, they do not envision the other individual's mind.

3. Self-recognition and mind reading

The fact that monkeys do not envision the other's mind when they predict another's behavior, however, does not necessarily imply that these animals are not intelligent. Level 2 of the process by which humans predict the behavior of others—that is, the assumption that the mechanism according to which others behave is the same as their own—may be a method that is specific to humans and great apes. One reason the study of a theory of mind involves the field of philosophy is that our theory of mind is not merely a theory of animal movement, like one of the theories that predict the movement of objects. This is because our theory of mind, or in other words, the attribution of mind to others,

Photo. 10–1 A male looking at his reflection in the mirror.

is closely related to the attribution of mind to oneself, that is, seeing others as a self-projection. In one of the previously described observations, the chimpanzee removed the grimace from his face with his fingers as if indicating that he understood his own state of mind and also that other individuals may understand his mind.

An experiment often presented on TV programs is one in which animals are shown a mirror. Great apes can certainly recognize their images in a mirror.(26)(27) G. Gallup showed that if a researcher put a mark on chimpanzees while they were sleeping, on a part of their body where they could not see it, for example, on their foreheads, the chimpanzees would raise their hands to their foreheads after seeing the mark in a mirror.

What happens if Tibetan monkeys are shown a mirror? I showed a mirror to the Tibetan monkeys at Huangshan. They did not seem to be able to understand that the image in the mirror was their own. I showed the adult male, Duyanlong, a mirror that was 20 cm×30 cm. Duyanlong was excited on seeing himself in the mirror and reached his hands behind the mirror (Photo. 10–1). He tried to touch the reflection of his own face in the mirror; he probably thought that it was another monkey. Then, I placed the mirror in front of the adult female, Hua. When she saw herself in the mirror, she was startled and stepped back. She, too, thought that there was another monkey behind the mirror. This behavior is similar to that of Japanese monkeys. However, their response to "the monkey in the mirror" was different. Male Japanese monkeys attempted to attack the monkey in the mirror and female Japanese monkeys usually ran away. However, Duyanlong smacked his lips and continued looking into the mirror. As soon as I showed the female, Hua, the mirror a second time, she became excited and approached in order to look into it; that is, she looked forward to seeing the monkey in the mirror.

I also brought the mirror to the infants who were playing. Initially, they were surprised, but after some time, all the infants curiously approached in order to look into the mirror. As several infants gathered in front of the mirror, they saw their own faces and those of the monkeys beside them. The infants alternated between looking at the faces of their friends in the mirror and their actual faces. They may have thought, "This individual is right here, but his "other self" has appeared in the mirror." In addition, they noticed that one individual who was not the "other selves" of their friends also appeared in the mirror. That individual was a stranger that they did not usually see. However, they did not appear to be able to understand that the strange individuals in the mirror were actually themselves.

Even some animals other than humans and great apes are able to use a mirror to look for hidden food. However, there are only a few reports of trained Japanese monkeys being able to recognize themselves in a mirror.[28][29]

Let us compare this ability of self-recognition to the ability of mind reading. Individuals belonging to species that could recognize themselves in a mirror, such as chimpanzees, could understand the minds of other individuals to some extent. Further, once self-recognition in a mirror was established, as a part of the developmental process of the individual, these chimpanzees came to understand the relationship between seeing and knowing. The chimpanzees who were unable to understand the relationship between seeing and knowing could not recognize themselves in the mirror.[30] Mind reading and self-recognition may have been closely related in not only developmental process of each individual but also an evolutionary process of the species. One can recognize oneself if one sees oneself relative to others in one's own mind. The development and evolution of the attribution of mind to self and the attribution of mind to others may be closely interlinked.

In order to understand the mind-reading ability of other animals, I believe that it is necessary to comprehensively study the various abilities that may be related to a theory of mind in humans (the ability to understand cause and effect relationships, including those in the physical world; the ability to understand analogies; the ability to imitate; the ability to have insights; and the ability of self-cognition), and the various types of social intelligence that each animal displays during interactions with others (the ability to predict the behavior of others and manipulate them, and understand that while others belong to the same species as themselves, they are not themselves). In addition to these problems, a significant area of future study will be to elucidate the way in which the attribution of mind to the self and to others are intertwined and acquired, on the basis of both of individual development and the evolution of species.

Furthermore, I would like to discuss one more issue. Studies on self-recognition have focused on whether or not an individual can recognize his or her image in a mirror. Similarly, the study of a theory of mind focuses on whether or not an individual has the ability to pass a false belief task; this is considered the minimum requirement for having a theory of mind. This prerequisite facilitates our understanding of a theory of mind. However, this task involves an evaluation of the mistaken beliefs of others. It is necessary to be careful because this task focuses on understanding the false knowledge of others. Considering the results obtained thus far, the dog, Koro, who I mentioned previously in an example, probably cannot understand that his owner does not know where the dog food is. However, Koro experiences feelings of joy and sadness, and, for example, he may be able to intuit that his owner is unhappy; this ability may be better than that of humans.

4. The Evolution of primates: How have they been thinking and living?

In Chapters 1 to 9, we have mainly discussed the lives of Tibetan monkeys in their social group. In addition, in Chapter 10, we have examined the social intelligence of primates, including humans and apes. Here, I would like to present my opinion on the evolution of social intelligence in primates. How have monkeys, apes, and humans thought, lived, and evolved?[31]

A long time ago, the common ancestors of primates were solitary, nocturnal, and insectivorous creatures. At that time, they may have already possessed the ability to identify other individuals of their own species, understand the relationships between particular individuals and themselves, and alter their own behavior according to their social relationships with others as well as the behavior of others.

Later, most primate species modified their lifestyle. They became primarily diurnal and ate parts of plants, and their body size increased. Significantly, they lived in stable groups for long periods of time. Such primates—anthropoids—may have understood not only the relationships between themselves and others but also those between other individuals, and may have manipulated others by acting in advance based on their predictions of the behavior of others.

The social intelligence of anthropoids may be related to a theory of mind, given that they evaluate themselves relative to others by distancing themselves from their own perspectives, and predict the behavior of others on the basis of an understanding of the differences between their own social relationships and those of others. However, with the exception of great apes and humans, primates do not employ a method to manipulate or predict the behavior of others by making assumptions regarding the mind and knowledge of other individuals.

Unlike other primate species, great apes and humans understand not only the differences between their own social relationships and those of others, but also the differences in the knowledge possessed by themselves and others. At the very least, some great apes understand the relationship between seeing and knowing, that is, they understand that seeing influences their knowledge. These abilities are more closely related to a theory of mind in humans than are the abilities of other anthropoids.

What is responsible for the evolution of these abilities among great apes and humans? Chimpanzees display the ability to have insights. When they utilize tools to obtain food, they understand the complicated processes and procedures involved in handling these tools. This ability to have insights, which allows them to predict the future, can be used in a decision-making process in a social context. In fact, it may be used with greater efficiency in a social context. Some individuals may think, "If I perform behavior A, my opponent will perform behavior P. Then, at that moment, I can perform behavior B. Therefore, I should first perform behavior A." These individuals determine their behaviors using a complicated process, and such individuals can behave in a manner advantageous to themselves in social interactions.

In addition, when the social environment, including the minds of individuals, became complex, it was necessary to change the method used to predict the behavior of others. There is one method used to predict this behavior, that is, the individual remembers each occurrence of the behavior in different situations and assumes that other individuals will perform the same behavior when they are in the same situations as those previously observed. However, once other individuals acquired a large amount of information and began to use a more complicated process to manage it, it was much more difficult to directly associate (1) the situation and (2) the behavior. It was no longer possible for the individual to think, "In conditions A, B, ..., and X, this individual will behave in this manner." In situations in which the process of decision-making used by other individuals is complex, it becomes necessary to associate (1) the situation of other individuals, (2) their state of mind, and (3) their behavior. In other words, we understand (1) the information that other individuals receive, (2) their decisions regarding how they process the information, and (3) their behavior. It is possible to predict the behavior of another individual by assuming that "in situation A, this individual sees X, in situation B, he possesses this knowledge, and in situation X, he has a certain preference, he thinks in a certain manner, and will then perform a certain behavior". If we do not attempt to understand the minds of others or why they behave the way they do, more information, memories, and the difficult process of supposition will be necessary. Although we need to determine which method is more efficient in various situations, imagining the decision-making process of other individuals with regard to the input and output of information offered greater advantages to great apes and humans. This favored the development of a theory of mind in humans. Once individuals employed this method to predict the behavior of others, they were probably conscious of their own decision-making process.

Moreover, after the ancestors of humans parted ways from the common ancestors of great apes and human, they acquired the ability to understand that the knowledge possessed by others was different from their own. Further, they came to understand a secondary belief—for example, "Mother thinks that Maxi believes the chocolate is on the green shelf"—which corresponds to an intention at level 3. Humans can efficiently convey information by teaching, precisely because the other individual does not possess this knowledge. The acquisition of linguistic abilities has further established this process. Deception came to be used for entertainment purposes as well. Further, humans performed altruistic behavior not only to ensure a reciprocation of the behavior in the future but also to please other individuals and share in their pleasure. A theory of mind in humans may have been acquired and evolved in this way, and may have eventually led to the development of human-specific mental abilities and social characteristics.

I disagree with the idea that humans are exceptional because they possess rational and other abilities that other animals do not possess. However, I certainly believe that they should cherish their unique abilities of mind reading, empathy, and sympathy.

However, great apes and humans, who employed specific methods of perceiving others as self-projections, were overtaken in the natural world by other groups of primates, that is, macaques and baboons, who predict the behavior of others without envisioning their minds. The group of great apes has been declining, at least in terms of the number of populations and the area in which they are distributed. Although humans have come to inhabit every part of the world and their population has grown to approximately 6, 000, 000, 000, this is an exceptional case.

On the other hand, the ancestor of the Tibetan monkeys separated from the ancestor of humans and great apes, diverged toward various macaque and baboon species after a certain period of time, and flourished in various areas in the world. They experienced many changes, but one of the important changes was the increase in the sizes of their social groups. The large size of their social groups caused the evolution of their social intelligence. When the number of members in the group increases, the total number of individual pairs in the group also increases rapidly. This increase in the size of the group meant that a larger number of individuals interacted with each other in the course of their daily lives and a larger number of social relationships, which these individuals must have understood, were formed. Thus, their social environment became more complex. Further, the individuals in the group not only lived together but were also simultaneously competitors for the resources and cooperators who formed alliances and defended the resources together. The fact that they were not always enemies increased the significance of the communication that they used to provide information to one another. On the other hand, the fact that they were not complete allies created a situation in which they obtained benefits by deceiving others by using their social intelligence.

In a more complex social environment, it is preferable for individuals to predict the behavior of others in the interactions in which they may be involved in the future by observing and memorizing the interactions among other individuals rather than memorizing the results of the interactions that they themselves are involved in. Among many species of primates, individuals of one sex emigrate from their natal group and join other groups. In this case, it is advantageous for these individuals to predict the behavior of others on the basis of observations, for example, predicting the behavior of the individuals in their new group.(32)

In addition, it is more efficient to predict the behavior of others by using the concept of social relationships. In a complex social environment, it is preferable for monkeys to categorize the relationship between certain individuals, for example, "A and B" and "P and Q," as a "mother-offspring relationship" or "a relationship in which A is dominant over B." This will enable them to realize that individuals in a certain type of relationship will behave in a particular manner. Monkeys can predict "A behaves in this manner toward B," if they understand "A and B is in a certain type of relationship" and remember "the individuals in this relationship behaved in this manner." This is a better method than an alternate method in which they memorize each interaction between the individuals in question, such as "A behaved in this manner toward B" or "P behaved in this manner toward Q." It is advisable for the monkeys to classify particularly limited situations such as "the competition over leftover corn at the provisioning site," or "the competition over the female monkey, Zhi" into larger categories or compare them to similar situations by using analogies.(33) As a result, anthropoids such as macaques evolved the ability to understand the social relationships among the members of their groups and form coalitions on the basis of this knowledge. These abilities have made the interactions and relationships among individuals more complex and have spurred the evolution of their social intelligence.

As we have seen thus far, (1) monkeys understand not only the relationships between themselves and others but also the relationships between other individuals. (2) They evaluate the relative fighting abilities of each individual that affect the results of agonistic interactions, and understand the balance of power in their relationships when they form alliances in agonistic interactions. (3) They manipulate others by predicting the behavior of others and modifying their own behavior accordingly. (4) An observational case indicates that monkeys may be able to understand that the perceptions of others are different from their own. However, when they predict the behavior of others, (5) monkeys do not make the effort to envision the minds of others. Instead, they memorize each individual's behavior in various situations, classify them into categories based on the relationships between the individuals and the social context, and predict the behavior of others on the basis of this classification. Thus, it can be held that monkeys such as Tibetan monkeys are not psychologists, but are skilled ethologists (behaviorists). As a reaearcher of social behavior, I have observed their daily social behavior of the Tibetan monkeys at Huangshan, China, and in this book, I have analyzed the various types of behaviors they engage in. However, even if I were to memorize all the results of the observations and join the group as another monkey, I would never be able to compete with Beihen and the other Tibetan monkeys in the Yulingkeng group. They are excellent observers of social behavior who can understand their society and the complex social environment they live in, and can play their cards well in various social interactions.

5. Another process in which huddling clusters are formed

I will conclude this book by reintroducing the interactions among the Tibetan monkeys at the sleeping site. I hope that the reader, who has, up to this point, been reading about the life of the Tibetan monkeys, will realize the meaning of their individual behavior and the social intelligence that drives it.

In this book, I have described various events—the upheaval in the ranks of the females (Chapter 7), young Duyanlong becoming the alpha male, and the practice infanticide (Chapter 8). The following section, however, will describe an event that preceded all these events. At this point, Huangmao was the alpha male of the Yulingkeng group, Duyanlong and Sanxiong were still juveniles, Baijamao was still alive and healthy, and Beihen would soon become four years old. Teng, the daughter of Zhi, gave birth to her first baby three days before this event, and Ye gave birth the day I made this observation.

The season was spring, and the date was April 14th, 1990. The time was 5:50 p.m. When I arrived at the sleeping site and looked at the monkeys through my binoculars, I found Duyanlong and Wushang sitting on a branch of a tree growing out of the side of the cliff. Duyanlong was grooming Wushang and holding Huagang, who would be one year old that spring. I surmised that Duyanlong had presented Huagang to Wushang and they had bridged successfully. Sanxiong, who was beside Duyanlong, appeared to want to join in the interaction between these two monkeys. When Duyanlong had finished grooming Wushang, he continued holding Huagang and sat down 1 m away from Wushang. Sanxiong approached Duyanlong as if to say, "I've been waiting for this moment." As soon as Duyanlong sat down, Sanxiong stood in front of him, pulled Huagang away from his belly, and bridged with Duyanlong. After the bridging was completed, Sanxiong forcibly took hold of Huagang, and the three monkeys sat down together. Three-year-old Chawei approached Wushang, who became alone, and groomed him.

The infants Baixiang, Zhiming, and Eshang were playing with Tougou near the sleeping site.

At 5:55 p.m., Chiyan and Yanbaiban, who were new members of the group during the fall of 1989, were seated near site A. By now, they seemed like real members of the Yulingkeng group. When Erxiong arrived at site A, Chiyan huddled with him.

Soon, Ye and Zhi joined this huddle, forming a huddle of four individuals at site A. Huangmao and Youshizhi were to their right, and Baijamao and Zhi joined them. Baijamao's daughter, Baixiang, who was almost one year old, approached her mother. Baijamao was grooming Zhi, whose daughter, Keai, had huddled with Ye. Teng approached Ye, this behavior was unusual. As I wondered about the reason for this behavior, I observed Teng and Ye bridging, using Ye's baby. Ye, had, in fact, given birth to a baby that day. I discovered that the newborn baby was a male—I saw his penis when he was turned over for bridging. Ye took hold of the baby's hind legs in order to turn him over, and moved her face closer to his penis. Ye later groomed Teng. Sanjiaolian and Keai competed with each other to observe this bridging.

Suddenly, Sanjiaolian bit Keai, who was seated beside her, and Keai ran away. Immediately after this incident, however, Sanjiaolian approached Keai and groomed her briefly. After Sanjiaolian had finished this grooming, Keai groomed her in return.

After grooming Sanjiaolian, Keai approached Erxiong. Erxiong groomed Keai for a very short time-he ran his fingers through her fur just once. Keai then groomed Erxiong.

Hua, too, approached the site. While Ye, who was holding her newborn baby, was grooming Teng, Sanjiaolian, who was right beside them, screamed at Teng. I would like the reader to note that the big fight that occurred among the females in December 1990 (described in Chapter 7) was also prompted by a quarrel between Sanjiaolian and Teng. These two individuals were already on bad terms in April 1990. Erxiong looked at them as though wondering what was happening. Hua, too, glared at Teng, who then ran away from the huddle. As it would be a disaster for Teng if her newborn baby was hurt, she stayed with her mother, Zhi.

The Tibetan monkeys continued to assemble at the sleeping site, and I started getting very busy with my observations.

Two young females, Beihen and Canglian, stayed together. Beihen approached Zhiming, held him, and groomed him. When these two individuals entered a cave, Canglian followed them in order to hold Zhiming. In reality, Canglian wanted to touch Teng's and Ye's newborn babies, but as the mothers would not allow her to do so, she had to make a compromise and settle for Zhiming, who had just become one year old and was mischievous.

Another pair of young females, Feimao and Tougou, were together. Eshang and Guer, too, joined them. Feimao and Tougou turned Guer over in order to use him for bridging.

At 6:24 p.m., the huddle comprising Zhi, Baijamao, Huangmao, and Youshizhi was dispersed, and Zhi approached site A. At that point, Ye, who was being groomed by Hua, and Erxiong and Canglian, were huddled at site A. After Zhi moved to site A, Teng occupied the place she had left and groomed Baijamao. Youshizhi followed Zhi, but on his way, he stopped to inspect Beihen's genitalia. At this point, Baijamao chased Beihen, who ran away. As a result, Teng and the alpha male, Huangmao, were the only remaining members of the huddle. As he had no other option, Huangmao, too, moved to site A. Hua welcomed Huangmao by performing the presenting behavior, and groomed him once he sat beside her. Baijamao stopped chasing Beihen and returned to site A.

Youshizhi took hold of Guer, approached Erxiong, and bridged with him. He then groomed Erxiong. While he was holding Guer, Youshizhi sometimes groomed both the infant and Erxiong. The huddling cluster that had been formed earlier was assimilated into the one at site A, which began to increase in size. Several infants joined this huddle in succession. Wushang, Chawei, Duyanlong, Huagang, and Sanxiong had formed another huddle and appeared likely to fall

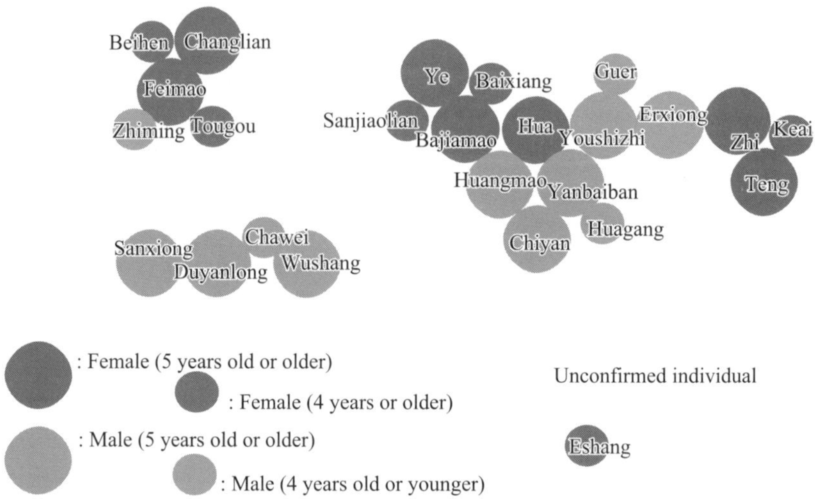

Fig. 10–1 Disposition of the huddles at the sleeping site on April 14, 1990.

asleep together. The young females, Beihen, Feimao, Tougou, and Canglian, and Zhiming, who was being held by Feimao, huddled together.

One monkey remained alone—Yanbaiban. Initially, he was with Chiyan and had been sitting in a corner this entire time. Although I believed that Yanbaiban had lost an opportunity to join a huddle, Huagang made the effort to extricate himself from Sanxiong's belly and approach Yanbaiban. Taking advantage of this opportunity, Yanbaiban immediately took hold of Huagang, approached Huangmao, and presented Huagang to Huangmao while smacking his lips. Huangmao received Huagang and bridged with Yanbaiban. After this, Yanbaiban remained beside Huangmao.

Zhi moved toward Erxiong, and her daughters, Teng and Keai, accompanied her. Ye looked a little nervous because she was holding her baby that had been born that day. She left the huddle and quietly moved to the other side. This resulted in a minor adjustment in the disposition of the huddle.

At 7:00 p.m., the following individuals were present at site A, seated from left to right: Ye, who had given birth to a baby that day, and Baijamao, who was flanked by her two daughters, Baixiang and Sanjiaolian. Baijamao, Huangmao, and Hua had formed a triangular huddle in the center of the larger huddle. Yanbaiban and Youshizhi huddled with that triangle. Yanbaiban was holding Huagang, and Youshizhi was holding Guer. Chiyan, who had moved away from Erxiong, reservedly huddled with Huangmao and Yanbaiban. After Erxiong had bridged with him, Youshizhi sat between Hua and Erxiong. Zhi, Teng, and Keai were beside Erxiong, on the other side. Beihen, Canglian, Tougou, and Feimao, who was holding Zhiming, had formed a huddle in the cave. Beside them, Wushang, Chawei, Duyanlong, and Sanxiong were asleep on a branch of a tree. The evening sun was slowly sinking behind the edge of Huangshan.

References

Chapter 1

1. Wada, K. (1996). *Chugoku sarugaku kikou: kouzan no yama ni kurasu (Japanese) <The study of monkeys in China: the study of their life on Mt. Huangshan>.* Tokai University Press.
2. Lindburg, D. G. (Ed.), (1980). *The Macaques: studies in ecology, behavior and evolution.* New York, Van Nostrand Reinhold.
3. Fooden, J. (1983). Taxonomy and evolution of the sinica group of macaques: 4. species account of *Macaca thibetana.* Fieldiana Zoology, 17: 1–20.
4. Wada, K., Xiong, C., & Wang, Q. (1987). On the distribution of Tibetan and rhesus monkeys in southern Anhui, China. Acta Theriologica, 7: 148–176.

Chapter 2

1. Humphrey, N. K. (1976). The social function of intellect. In P. P. G. Bateson & R. A. Hinde (Eds.), *Growing points in ethology*, Cambridge, Cambridge University Press, pp. 303–317.
2. Jolly, A. (1966). Lemur social behavior and primate intelligence. Science, 153, 501–506.
3. Byrne, R., & Whiten, A. (Eds.), (1988). *Machiavellian intelligence: social expertise and the evolution of intellect in monkeys, apes, and humans.* Oxford, Oxford University Press.
4. Sawaguchi, T., & Kudo, H. (1990). Neocortical development and social structure in primates. Primates, 31: 283–289.
5. Wrangham, R. W. (1980). An ecological model of female-bonded primate groups. Behaviour, 75: 262–300.
6. Axelrod, R., & Dion, D. (1982). The evolution of cooperation. Science, 211: 1390–1396.
7. Nowak, M., & Sigmund, K. (1993). A strategy of win-stay, lose-shift that outforms tit-for-tat in the prisoner's dilemma game. Nature, 364: 56–58.
8. Ogawa, H., Matsumura, S., Oura, H., Nakmura, M., Okamoto, K., & Yabuta, S. (1998). Shakaikoudou to gemu riron (Japanese) <Social behavior and game theory>. Primate Research, 14 (3): 209–221.
9. Fushimi, T. (1990). A functional analysis of another individual's behavior as discriminative stimulus for a monkey. Journal of the Experimental Analy-

sis of Behavior, 53: 285–291.
10. Hauser, M. D., & Nelson, D. (1991). Intentional signaling in animal communication. Trends of Ecology & Evolution, 6: 186–189.
11. Krebs J. R., & Dawkins R. (1984). Animal signals: mindreading and manipulation. In J. R. Krebs & N. B. Davis (Eds.), *Behavioural ecology: an evolutionary approach, 2nd edn.*, Oxford, Blackwell Scientific Publications, pp. 380–402.
12. Ogawa, H. (1996). Social interactions among a mother, an adopted infant, and a biological infant of Japanese macaques in the Arashiyama F troop (Japanese with English abstract). Primate Research, 12 (1): 1–10.
13. Ogawa, H. (1998). Adoption and social interactions among a mother and 'twin' offspring in *Macaca fuscata*. Folia Primatologica, 69 (2): 100–105.
14. Matsuzawa, T. (1991). *Chimpanzee kara mita sekai (Japanese) <The world recognized by chimpanzees>*. Tokyo University Press.

Chapter 3

1. Hall, E. T. (1966). *The hidden dimension.* New York, Doubleday.
2. Altmann, J. (1974). Observational study of behaviour: sampling methods. Behaviour, 49: 227–265.
3. Martin, P., & Bateson, P. (1993). *Measuring behaviour: an introductory guide, 2nd ed.* Cambridge University Press.

Chapter 4

1. Sugiyama, Y. (1971). Characteristics of the social life of bonnet macaques (*Macaca radiata*). Primates, 12: 247–266.
2. Deag, J. M., & Crook, J. H. (1971). Social behaviour and "agonistic buffering" in the wild Barbary macaque, *Macaca sylvana* L. Folia Primatologica, 15: 183–200.
3. Taub, D. M. (1980). Testing the "agonistic buffering" hypothesis. Behavioral Ecology and Sociobiology, 6: 187–197.
4. Dunbar, R. I. (1984). Infant-use by male gelada in agonistic contexts: agonistic buffering, progeny protection or soliciting support? Primates, 25: 28–35.
5. Packer, C. (1980). Male care and exploitation of infants in *Papio anubis*. Animal Behaviour, 28: 512–520.
6. Estrada, A. (1977). Social relationships in a free-ranging troop of stumptail macaques (*Macaca arctoides*): male-care behaviour 1. Primates, 18: 793 – 813.
7. Itani, J. (1959). Paternal care in the wild Japanese monkey, *Macaca fuscata fuscata*. Primates, 2: 61–93.
8. Silk, J. B., & Samuels, A. (1984). Triadic interactions among *Macaca*

radiata: passports and buffers. American Journal of Primatology, 6: 373–376.

9. Taub, D. M. (1984). Male caretaking behavior among wild Barbary macaques (*Macaca sylvanus*). In Taub, D. M. (Ed.), *Primate paternalism*, New York, Van Nostrand Reinhold, pp. 20–55.
10. Ogawa, H. (1995). Recognition of social relationships in bridging behavior among Tibetan macaques (*Macaca thibetana*). American Journal of Primatology, 35: 305–310.
11. Cheney, D. L., & Seyfarth, R. M. (1990). *How monkey see the world.* University of Chicago Press.
12. de Waal, F. B. M. (1989). *Peace making among primates.* Cambridge, Harvard University Press.
13. Dasser, V. (1988). Social concept in Java monkey. Animal Behaviour, 36: 225–230.
14. Harcourt, A. H. (1988). Alliances in contests and social intelligence. In R. Byrne & A. Whiten (Eds.), *Machiavellian intelligence: social expertise and the evolution of intellect in monkeys, apes, and humans*, Oxford, Oxford University Press, pp. 132–152.

Chapter 5

1. Zhao, Q. (1993). Sexual behavior of Tibetan macaques at Mt. Emei, China. Primates, 34: 431–444.
2. Zhao, Q. (1994). Mating composition and intergroup transfer of males in Tibetan macaques (*Macaca thibetana*) at Mt. Emei, China. Primates, 35: 57–68.
3. Zhao, Q., & Deng, Z. (1988). *Macaca thibetana* at Mt. Emei, China: 1 a cross-sectional study of growth and development. American Journal of Primatology, 16: 251–260.
4. Zhao, Q., & Deng, Z. (1988). *Macaca thibetana* at Mt. Emei, China: 2 birth seasonality. American Journal of Primatology, 16: 261–268.
5. Zhao, Q., & Deng, Z. (1988). *Macaca thibetana* at Mt. Emei, China: 3 group composition. American Journal of Primatology, 16: 269–273.

Chapter 6

1. Xiong, C. (1989). Tibetan monkey no osu no koubikoudou (Japanese) <Mating behavior of male Tibetan macaques>. Monkey, 227–228: 20–28.
2. Whitten, P. L. (1987). Infants and adult males. In B. Smuts, D. Cheney, R. Seyfarth, R. Wrangham, & T. Struhsaker (Eds.), *Primate Societies*, Chicago, University of Chicago Press, pp. 343–357.
3. van Schaik, C. P., & Paul, A. (1997). Male care on primates: does it reflect

paternity? Evolutionary Anthropology, 5 (5): 152–156.

4. Kuester, J., & Paul, A. (1996). Infant handling by Barbary macaques (*Macaca sylvanus*) at Affenberg Salem: testing functional and evolutionary hypotheses. Behavioral Ecology and Sociobiology, 39: 133–145.
5. Paul, A., & Kuester, J. (1994). Kinship, familiarity and mating avoidance in Barbary macaques, *Macaca sylvanus*. Animal Beahviour, 48: 1183–1194.
6. Smuts, B. B. (1985). *Sex and friendship in baboons.* Aldine Publishing Company.
7. Dunbar, R. I. (1984). Infant-use by male gelada in agonistic contexts: agonistic buffering, progeny protection or soliciting support? Primates, 25: 28–35.
8. Ogawa, H. (1995). Bridging behavior and other affiliative interactions among male Tibetan macaques (*Macaca thibetana*). International Journal of Primatology, 16 (5): 707–729.
9. Ogawa, H. (1995). Triadic male-female-infant relationships and bridging behaviour among Tibetan macaques (*Macaca thibetana*). Folia Primatologica, 64 (3): 153–157.
10. Premack, D. (1983). Animal cognition. Annual Review of Psychology, 34: 351–362.
11. Wada, K., & Xiong, C. (1996). Population changes of Tibetan monkeys with special regard to birth interval. In T. Shiotake & K. Wada (Eds.), *Variations in the Asian macaques*, Tokai University Press.

Chapter 7

1. Ogawa, H. (1991). Tibetan monkey no mesutachi (Japanese) <Social relationships among female Tibetan macaques>. Monkey, Vol. 35 (3 • 4), No. 239 • 240: 30–35.
2. Ogawa, H. (1993). Sonogono Tibetan monkey no mesutachi (Japanese) <Social relationships among Tibetan macaques 2>. Monkey, Vol. 36 (3 • 4), No. 245 • 246: 17–21.
3. Muroyama, Y. (1991). Mutual reciprocity of grooming in female Japanese macaques (*Macaca fuscata*). Behaviour, 119: 161–170.
4. Li, J., & Wang, Q. (1994). Allogrooming in wild Tibetan macaques (*Macaca thibetana*), with special reference to social structure. In J. J. Roeder (Ed.), *Current Primatology: Vol. 2, social development, learning and behaviour*, pp. 185–191.
5. Harcourt, A. H. (1992). Coalitions and alliances: are primates more complex than other nonprimates? In A. H. Harcourt & F. B. M. de Waal (Eds.), *Coalitions and alliances in humans and other animals*, New York, Oxford University Press, pp. 445–472.
6. Noë, R. A. (1990). Veto game played by baboons: a challenge to use of the

prisoner's dilemma as a paradigm for reciprocity and cooperation. Animal Behaviour, 39, 78–90.

7. de Waal, F. B. M. (1982). Chimpanzee politics. London, Jonathan Cape.
8. Nishida, T. (1983). Alpha status and agonistic alliance in wild chimpanzees (*Pan troglodytes*). Primates, 24: 318–336.
9. de Waal, F. B. M., & Luttrell, L. M. (1989). Toward a comparative socioecology of the genus *Macaca*: different dominance styles in rhesus and stumptail monkeys. American Journal of Primatology, 19: 83–109.
10. van Schaik, C. P. (1989). The ecology of social relationships amongst female primates. In V. Stein & R. Foley (Eds.), *Comparative socioecology: the behavioural ecology of humans and other mammals*, Oxford, Blackwell Scientific Publications, pp. 195–218.
11. Matsumura, S., & Kobayashi, T. (1998). A game model for dominance relations among group-living animals. Behavioral Ecology and Sociobiology, 42: 77–84.

Chapter 8

1. Sugiyama, Y. (1980). *Kogoroshi no Koudougaku (Japanese) <Ethology of infanticide>*. Hokuto Press.
2. Hasegawa, M. (1992). Reichourui no kogoroshi o meguru shomondai (Japanese) <Review of infanticide in primates>. In Y. Ito (Ed.), *Doubutsu shakai ni okeru kyoudou to kougeki (Japanese) <Cooperation and aggression in animal groups>*, Tokai University Press, pp. 185–222.

Chapter 9

1. Wada, K., Tokida, E., & Ogawa, H. (in preparation). The influence of snowfall, temperature, and social relationships on sleeping clustrers of Japanese monkeys during winter at Shiga Heights.
2. Anderson, J. R., & McGrew, W. C. (1984). Guinea baboons (*Papio papio*) at sleeping site. American Journal of Primatology, 6: 1–14.
3. Takahashi, H. (1997). Huddling relationships in night sleeping groups among wild Japanese macaques in Kinkazan Island during winter. Primates, 38: 57–68.

Chapter 10

1. Byrne, R. (1995). *Thinking ape.* Oxford, Oxford University Press.
2. Kummer, H. (1982). Social knowledge in free-range primates. In D. R. Griffin (Ed.), *Animal mind-human mind*, Berlin, Springer-Verlag, pp. 113–130.
3. de Waal, F. B. M. (1982). *Chimpanzee politics.* London, Jonathan Cape.
4. Menzel, E. W. (1974). A group of chimpanzees in a 1-acre field: leadership

and communication. In A. M. Schrier & F. Stollniz (Eds.), *Behavior of non-human primates*, New York, Academic Press, pp. 83–153.

5. Matsuzawa, T. (1991). *Chimpanzee mind (Japanese) < Chimpanzee mind>*. Iwanami Shoten.
6. Maynard-Smith, J., & Packer, G. A. (1976). The logic of asymmetric contests. Animal Behaviour, 24: 159–175.
7. Krebs J. R., & Dawkins, R. (1984). Animal signals: mindreading and manipulation. In J. R. Krebs & N. B. Davis (Eds.), *Behavioural ecology: an evolutionary approach, 2nd edn.*, Oxford, Blackwell Scientific Publications, pp. 380–402.
8. Grafen, A. (1990). Biological signals as handicaps. Journal of Theoretical Biology, 144: 517–546.
9. Munn, C. A. (1986). Birds that "cry wolf." Nature, 319, 143–145.
10. Hauser, M. D. (1992). Cost of deception: cheaters are punished in rhesus monkeys (*Macaca mulatta*). Proceedings of National Academy of Sciences of the United States of America, 89: 12137–12139.
11. Cheney, D. L., & Seyfarth, R. M. (1988). Assessment of meaning and the detection of unreliable signals by vervet monkeys. Animal Behaviour, 36, 477–486.
12. Wiley, R. H. (1994). Errors, exaggeration, and deception in animal communication. In L. A. Real (Ed.), *Behavioral mechanisms in evolutionary biology*, Chicago, The University of Chicago Press, pp. 157–189.
13. Premack, D., & Woodruff, G. (1978). Does the chimpanzee have a theory of mind? The Behavioral and Brain Sciences, 4: 515–526.
14. Wellman, H. M. (1991). From desires to beliefs: acquisition of a theory of mind. In A. Whiten (Ed.), *Natural theories of mind: evolution, development and simulation of everyday mindreading*, Basil Blackwell, pp. 19–38.
15. Wimmer, H., & Perner, J. (1983). Beliefs on beliefs: representation and constructiong function of wrong beliefs in young children's understanding of deception. Cognition, 13: 103–128.
16. Premack, D. (1988). Does the chimpanzee have a theory of mind? revisited. In R. Byrne & A. Whiten (Eds.), *Machiavellian intelligence: social expertise and the evolution of intellect in monkeys, apes, and humans*, Oxford, Oxford University Press, pp. 160–179.
17. Povinelli, D. J., Nelson, K. E., & Boysen, S. T. (1990). Inferences about guessing and knowing by chimpanzees (*Pan troglodytes*). Journal of Comparative Psychology, 107: 203–210.
18. Dennett, D. C. (1983). Intentional systems in cognitive ethology: the "anglossian paradigm" defended. The Behavioral and Brain Sciences, 6: 343 – 390.

19. Mitchell, R. W., & Thompson, N. S. (Eds.), (1986). *Deception: perspectives on human and nonhuman deceit.* New York, University of New York Press.
20. Skutch, A. F. (1976). *Parental birds and their young.* Austin, Texas, University of Texas Press.
21. Povinelli, D. J., Parks, K. A., & Novak, M. A. (1991). Do rhesus monkeys (*Macaca mulatta*) attribute knowledge and ignorance to others? Journal of Comparative Psychology, 105: 318–324.
22. Itakura S. (1997). Understanding of the human mind by nonhuman primates: the theory of mind in nonhuman primates. Japanese Psycological Review, 40 (1): 8–21.
23. Byrne, R., & Whiten, A. (1990). Tactical deception in primates: the 1990 database. Primate Report, 27: 1–101.
24. Caro, T. M., & Hauser, M. D. (1992). Is there teaching in non-human animals? Quarterly Review of Biology, 67: 151–174.
25. Cheney, D. L., & Seyfarth, R. M. (1990). Attending to behaviour versus attending to knowledge: examining monkeys' attribution of mental states. Animal Behaviour, 40: 742–753.
26. Gallup, G. G. (1970). Chimpanzees: self cognition. Science, 167: 86–87.
27. Inoue, T. (1994). Mirror self-recognition among infant chimpanzees: application of longitudinal and cross-sectional methods. The Japanese Journal of Developmental Psycology, 5: 51–60.
28. Boccia, M. (1994). Mirror behavior in macaques. In S. T. Parker, R. W. Mitchell, & M. L. Boccia (Eds.), *Self-awareness in animals and humans: developmental perspectives*, Cambridge University Press.
29. Itakura, S. (1987). Use of a mirror to direct their responses in Japanese monkeys (*Macaca fuscata fuscata*). Primates, 28: 343–352.
30. Povinelli, D. J., Rulf, A. B., & Bierschwale, D. T. (1994). Absence of knowledge attribution and self recognition in young chimpanzees (*Pan troglodytes*). Journal of Comparative Psychology, 108: 74–80.
31. Ogawa, H. (1997). Intraspecies theory of mind in primates: the evolution of social intelligence. Japanese Psycological Review, 40 (1): 22–38.
32. Seyfarth, R. M., & Cheney, D. L. (1994). The evolution of social cognition in primates. In L. A. Real (Ed.), Behavioral mechanisms in evolutionary biology, Chicago, The University of Chicago Press, pp. 371–389.
33. Kummer, H. (1982). Social knowledge in free-range primates. In D. R. Griffin (Ed.), *Animal mind-Human mind*, Berlin, Springer-Verlag, pp. 113–130.

Afterword for the English version

1. Ogawa, H. & Takahashi, H. (2003). Triadic positions of Tibetan macaques huddling at a sleeping site. International Journal of Primatology, 24 (3):

591–606.
2 . Li, J., Wang, Q. and Han, D. (1996). Fission in a free-ranging Tibetan macaque group at Huangshan Mountain, China. Chinese Sci. Bull, 41: 1377 – 1381.
3 . Berman, C. M. & Li, J. (2002). Impact of translocation, provisioning and range restriction on a group of *Macaca thibetana*. International Journal of Primatology, 23: 283–397.
4 . Berman, C. M., Ionica, C. S. and Li, J. (2004). Dominance style among *Macaca thibetana* on Mt. Huangshan, China. Internatioanl Journal of Primatology, 25: 1283–1312.

Remark: In Japanese articles, we translated the Japanese title to English and add the English title in < >. Therfore, the authors of this book, Ogawa and Yanagi, are responsible for it if there is any unsuitable expression in the English titles. In Japanese articles with English title and abstract, we show the English title and add "Japanese with English abstract" in ( ).

Afterword

First of all, I would like to show my appreciation to you for reading this book in spite of the weird and suspicious title, "Wily Monkeys: Social Intelligence of Tibetan Macaques."

The topic of this book is about social relationships among Tibetan monkeys at Huangshan, where I stayed several times over the years from 1985 to 1993.

As I introduced in chapter 1, the beginning of my study was so rash that I decided to start this study because I was simply attracted to the name, 'Tibetan' monkeys. Actually I cannot get out of my habit of making instant decisions, so I ended up going to Africa later. I asked Dr. Masazumi Mitani at the Museum of Nature and Human Activities, Hyogo to take me to Nouabale-Ndoki National Park, and I visited Republic of Congo in 1994. In addition, when Dr. Takayoshi Kano at Primate Research Institute, Kyoto University, asked me if I was interested in conducting research in Tanzania, I immediately answered "Yes, I will do it!" once again. That is why I have been making my study on wild chimpanzees in the savanna woodland in Tanzania since 1994. Since then I have not conducted any fieldwork for Tibetan monkeys, so I feel sorry for researchers who expect to know more about these monkeys. During my recent study on chimpanzees in the Ugalla area, Tanzania, I could rarely find chimpanzees even after I walked in the savanna woodland and looked for chimpanzees all day. This often brings back my good memories at Huangshan in China. At Huangshan, my observational condition was so great that I could closely observe the social behavior of Tibetan monkeys all day. I wrote this book out of such sentimental feelings.

My attempt to reveal the social intelligence of Tibetan monkeys in their complex social interactions involving three and more individuals digressed many times to the description about their interesting social behavior. If you found this book interesting, it is because of the Tibetan monkeys who are even more interesting when you actually see them. If you found this book boring, it is because I did not do a well enough job describing their interesting behavior. In this book, I also write about the various episodes that happened to the Tibetan monkeys during my observations. I even dared to write about alliances among females and huddling clusters, which I am not yet confident in the analyses and so may potentially be a wrong idea. It is my honor to have this book added to publications of Kyoto University Press. (The Japanese version of this book was published as Vol. 7 of the 'Ecological Science library.')

Dr. Kazuo Wada, who was at Primate Research Institute, Kyoto University when I started my study at Huangshan, has given me much support and guidance since the beginning of my study. I would not have had a chance to meet Tibetan monkeys and this book would have never been published without his dedication, so I would like to give a special acknowledgement to him. I would also like to express my appreciation to Prof. Qishan Wang at Anhui University and Mr. Chenpei Xiong, Mr. Jinhua Li, and Mr. Ming Li of the Huangshan Primate Research team for their cooperation with my study. Many thanks to the staff at Anhui University who took care of my administrative work, such as my permission to stay and study at Huangshan. I would like to show my appreciation to the people of Fuxi village, especially Mr. Xinming Cheng, for supporting my study over there. Without their help, this project would have not been possible.

Most of my data analyses were conducted when I was a graduate student at the Primate Research Institute of Kyoto University. Participants in the seminar on Ecology and Social Behavior of Primates at Primate Research Institute, especially, the chief professor of the department of Social Behavior, Dr. Takayoshi Kano, provided me with valuable opinions and great encouragement. Yasuyuki Muroyama spared a large amount of time for my study and gave me many pieces of advice. In the process of writing my dissertation, I gained valuable opinions from Drs. Yukimaru Sugiyama, Akio Mori, and Tetsuro Matsuzawa. Dr. Tetsuro Matsuzawa, who introduced me to Kazuo Wada, helped me a lot in this process too.

For my study on Japanese monkeys in Arashiyama, I had much support from the staff at Iwatayama Natural Park (currently named Arashiyama Monkey Park, Iwatayama). Many thanks to Dr. Jun'ichiro Itani for attending the presentation of my graduation thesis. It was his last year at Kyoto University. Many thanks to Dr. Yukio Takahata, who largely influenced me in the beginning of my life as a researcher.

Dr. Shuichi Matsumura, who has studied Japanese macaques and other macaques, has helped me realize many characteristics of the social behavior and the social relationships of Tibetan monkey while we talked about macaques for the past 10 years. Dr. Hiroyuki Takahashi gave me his cooperation with my analysis of the huddling clusters.

I am also appreciative of Dr. Toshisada Nishida for introducing me to Kyoto University Press. In addition to my appreciation for Mr. Testuya Suzuki and Mr. Shigekazu Takagaki at Kyoto University Press, I would rather like to apologize to them for extending the deadline many times.

I greatly appreciate my parents, Toshihiko Ogawa and Hiromi Ogawa, who worry about me every time I go away for my fieldwork, and my wife, Orie Ogawa. She understands me in a way no one else can, for example when I used to ask such a strange question, as "How come is the group, Dreams Come True, doing so well, although it consists of one woman and two men".

Finally, I would like to appreciate and dedicate this book to the Tibetan monkeys in the Yulingkeng group: Beihen, Baijamao, Zhi, Huangmao, Wushang, Duyanlong, and Sanxiong, for not only providing me many significant inspirations and materials to study, but also making my research life in Fuxi village at Huanshang precious.

On July 14, 1999
Hideshi Ogawa

Afterword for the English version

This book was originally published in Japanese in 1999 by Kyoto University Press. I would like to describe the reason that this English version is being published and what has been going on in the Tibetan monkeys at Huangshan since I finished my observation in 1993.

As I mentioned in the afterword, I started a new study, an ecological survey of chimpanzees in Tanzania, in 1994. Dr. Kazuo Wada also started observations of the golden snub-nosed monkeys in China. So he and I did not collect behavioral data on the Tibetan monkeys after 1993 though I continued data analyses[(1)]. On the other hand, however, Dr. Jinhua Li, who is currently a professor at Anhui University, has been keeping observation and identification of the Tibetan monkeys and recording the changes in composition of the Yulingkeng group. In addition, Dr. Carol Berman, a professor of the Department of Anthropology, the State University of New York at Buffalo, became interested in Tibetan monkeys and stayed at Huangshan to observe them. When I attended the 15th Congress of the International Primatological Society in 2002 at Beijing, China, I met C. Berman and gave her the original Japanese version of this book. Later, she asked Ms. Akie Yanagi, who is a graduate student of her department, to translate it to English. When I was a visiting scholar of University of California, San Diego during the sabbatical in 2004, I revised the manuscript translated by A. Yanagi. Then, we finally completed the English version, and Kyoto University Press and Trans Pacific Press published it with the financial support of Chukyo University Fund for Research and Publication.

When I visited China in 2002, I also stayed at Fuxi village for a few days and met the Tibetan monkeys at Huangshan after a 9 year interval. I saw some monkeys whom I had observed in 1989–1993. I saw that Beihen and Tougou were grooming each other. They were 3-year-old in 1989 and were full adult females then. I could not meet Duyanlong because he emigrated from the Yulingkeng group. In stead, I was surprised that Huagang, who was one-year-old in 1989, became a big adult male. I was also surprised that many tourists climbed up to Fuxi and observed the monkeys. People in Fuxi village and the local government carried out the plan that they would make a park where tourists could see monkeys. They built a viewing pavilion, made "Huangshan Monkey Center" and started the tourism in 1994. While the Tibetan monkeys were provisioned for only during our observational periods before 1993, they are now provisioned for every day throughout the year. Thanks to the suggestion of such by J. Li, however, the park staff made a rule, which does not allow tourists to feed or touch the monkeys. Unlike Japanese monkeys in provisioned groups in Japan, the population of the Yulingkeng group did not increase because of group fissions and many cases of infant death.

Up to now, J. Li and C. Berman found many interesting things, some of which are listed in the references[(2) (3) (4)]. Further infanticides unfortunately occurred in the Yulingkeng group. This may be related with frequent aggression in the small restricted provisioning area for tourism. Social relationships among female Tibetan monkeys may have different aspects from my impression that I obtained from the rank reversal among females in 1990. Although I am not concerned with the management of the park and observation of the Tibetan monkeys in the park now, I would like to contribute to the study of Tibetan monkeys by presenting comparable data to that of theirs. In addition, Mr. Yin Huabao, a graduate student of Anhui University, and Mr. Consuel Severus Inocia, a graduate student of State University of New York, Buffalo joined the study of Tibetan monkeys at Huangshan. I believe that they will make progress with the studies of Tibetan monkeys and the Tibetan monkeys give them further opportunities for them.

As well as the studies of the Tibetan monkeys, a lot of studies have progressed in primatology during these 10 years. I could not reflect those recent findings in this book when I revised the original manuscript. Also, I omit most of the methods and results of the statistical tests from this book and I write the results only when they were not significant. Please refer to my articles in scientific journals for the detailed methods and results. In this sense, I am not fully satisfied with this book. In this book, however, I introduced many episodes of the Tibetan monkeys and many ideas of mine, which I could not include in other scientific articles. I am very happy that I could publish this book in English. I hope that a large number of people read this book, and come away from it with an understanding of how attractive the Tibetan monkeys are themselves, and to study.

On December 17, 2005
Hideshi Ogawa

Index